FREE Test Taking Tips DVD Offer

To help us better serve you, we have developed a Test Taking Tips DVD that we would like to give you for FREE. **This DVD covers world-class test taking tips that you can use to be even more successful when you are taking your test.**

All that we ask is that you email us your feedback about your study guide. Please let us know what you thought about it – whether that is good, bad or indifferent.

To get your **FREE Test Taking Tips DVD**, email freedvd@studyguideteam.com with "FREE DVD" in the subject line and the following information in the body of the email:

 a. The title of your study guide.

 b. Your product rating on a scale of 1-5, with 5 being the highest rating.

 c. Your feedback about the study guide. What did you think of it?

 d. Your full name and shipping address to send your free DVD.

If you have any questions or concerns, please don't hesitate to contact us at freedvd@studyguideteam.com.

Thanks again!

SHSAT Prep Questions 2020 & 2021

Three SHSAT Practice Tests for the Specialized High School Admissions Test [3rd Edition]

Test Prep Books

Table of Contents

Quick Overview

As you draw closer to taking your exam, effective preparation becomes more and more important. Thankfully, you have this study guide to help you get ready. Use this guide to help keep your studying on track and refer to it often.

This study guide contains several key sections that will help you be successful on your exam. The guide contains tips for what you should do the night before and the day of the test. Also included are test-taking tips. Knowing the right information is not always enough. Many well-prepared test takers struggle with exams. These tips will help equip you to accurately read, assess, and answer test questions.

A large part of the guide is devoted to showing you what content to expect on the exam and to helping you better understand that content. In this guide are practice test questions so that you can see how well you have grasped the content. Then, answer explanations are provided so that you can understand why you missed certain questions.

Don't try to cram the night before you take your exam. This is not a wise strategy for a few reasons. First, your retention of the information will be low. Your time would be better used by reviewing information you already know rather than trying to learn a lot of new information. Second, you will likely become stressed as you try to gain a large amount of knowledge in a short amount of time. Third, you will be depriving yourself of sleep. So be sure to go to bed at a reasonable time the night before. Being well-rested helps you focus and remain calm.

Be sure to eat a substantial breakfast the morning of the exam. If you are taking the exam in the afternoon, be sure to have a good lunch as well. Being hungry is distracting and can make it difficult to focus. You have hopefully spent lots of time preparing for the exam. Don't let an empty stomach get in the way of success!

When travelling to the testing center, leave earlier than needed. That way, you have a buffer in case you experience any delays. This will help you remain calm and will keep you from missing your appointment time at the testing center.

Be sure to pace yourself during the exam. Don't try to rush through the exam. There is no need to risk performing poorly on the exam just so you can leave the testing center early. Allow yourself to use all of the allotted time if needed.

Remain positive while taking the exam even if you feel like you are performing poorly. Thinking about the content you should have mastered will not help you perform better on the exam.

Once the exam is complete, take some time to relax. Even if you feel that you need to take the exam again, you will be well served by some down time before you begin studying again. It's often easier to convince yourself to study if you know that it will come with a reward!

Test-Taking Strategies

1. Predicting the Answer

When you feel confident in your preparation for a multiple-choice test, try predicting the answer before reading the answer choices. This is especially useful on questions that test objective factual knowledge. By predicting the answer before reading the available choices, you eliminate the possibility that you will be distracted or led astray by an incorrect answer choice. You will feel more confident in your selection if you read the question, predict the answer, and then find your prediction among the answer choices. After using this strategy, be sure to still read all of the answer choices carefully and completely. If you feel unprepared, you should not attempt to predict the answers. This would be a waste of time and an opportunity for your mind to wander in the wrong direction.

2. Reading the Whole Question

Too often, test takers scan a multiple-choice question, recognize a few familiar words, and immediately jump to the answer choices. Test authors are aware of this common impatience, and they will sometimes prey upon it. For instance, a test author might subtly turn the question into a negative, or he or she might redirect the focus of the question right at the end. The only way to avoid falling into these traps is to read the entirety of the question carefully before reading the answer choices.

3. Looking for Wrong Answers

Long and complicated multiple-choice questions can be intimidating. One way to simplify a difficult multiple-choice question is to eliminate all of the answer choices that are clearly wrong. In most sets of answers, there will be at least one selection that can be dismissed right away. If the test is administered on paper, the test taker could draw a line through it to indicate that it may be ignored; otherwise, the test taker will have to perform this operation mentally or on scratch paper. In either case, once the obviously incorrect answers have been eliminated, the remaining choices may be considered. Sometimes identifying the clearly wrong answers will give the test taker some information about the correct answer. For instance, if one of the remaining answer choices is a direct opposite of one of the eliminated answer choices, it may well be the correct answer. The opposite of obviously wrong is obviously right! Of course, this is not always the case. Some answers are obviously incorrect simply because they are irrelevant to the question being asked. Still, identifying and eliminating some incorrect answer choices is a good way to simplify a multiple-choice question.

4. Don't Overanalyze

Anxious test takers often overanalyze questions. When you are nervous, your brain will often run wild, causing you to make associations and discover clues that don't actually exist. If you feel that this may be a problem for you, do whatever you can to slow down during the test. Try taking a deep breath or counting to ten. As you read and consider the question, restrict yourself to the particular words used by the author. Avoid thought tangents about what the author *really* meant, or what he or she was *trying* to say. The only things that matter on a multiple-choice test are the words that are actually in the question. You must avoid reading too much into a multiple-choice question, or supposing that the writer meant something other than what he or she wrote.

5. No Need for Panic

It is wise to learn as many strategies as possible before taking a multiple-choice test, but it is likely that you will come across a few questions for which you simply don't know the answer. In this situation, avoid panicking. Because most multiple-choice tests include dozens of questions, the relative value of a single wrong answer is small. As much as possible, you should compartmentalize each question on a multiple-choice test. In other words, you should not allow your feelings about one question to affect your success on the others. When you find a question that you either don't understand or don't know how to answer, just take a deep breath and do your best. Read the entire question slowly and carefully. Try rephrasing the question a couple of different ways. Then, read all of the answer choices carefully. After eliminating obviously wrong answers, make a selection and move on to the next question.

6. Confusing Answer Choices

When working on a difficult multiple-choice question, there may be a tendency to focus on the answer choices that are the easiest to understand. Many people, whether consciously or not, gravitate to the answer choices that require the least concentration, knowledge, and memory. This is a mistake. When you come across an answer choice that is confusing, you should give it extra attention. A question might be confusing because you do not know the subject matter to which it refers. If this is the case, don't eliminate the answer before you have affirmatively settled on another. When you come across an answer choice of this type, set it aside as you look at the remaining choices. If you can confidently assert that one of the other choices is correct, you can leave the confusing answer aside. Otherwise, you will need to take a moment to try to better understand the confusing answer choice. Rephrasing is one way to tease out the sense of a confusing answer choice.

7. Your First Instinct

Many people struggle with multiple-choice tests because they overthink the questions. If you have studied sufficiently for the test, you should be prepared to trust your first instinct once you have carefully and completely read the question and all of the answer choices. There is a great deal of research suggesting that the mind can come to the correct conclusion very quickly once it has obtained all of the relevant information. At times, it may seem to you as if your intuition is working faster even than your reasoning mind. This may in fact be true. The knowledge you obtain while studying may be retrieved from your subconscious before you have a chance to work out the associations that support it. Verify your instinct by working out the reasons that it should be trusted.

8. Key Words

Many test takers struggle with multiple-choice questions because they have poor reading comprehension skills. Quickly reading and understanding a multiple-choice question requires a mixture of skill and experience. To help with this, try jotting down a few key words and phrases on a piece of scrap paper. Doing this concentrates the process of reading and forces the mind to weigh the relative importance of the question's parts. In selecting words and phrases to write down, the test taker thinks about the question more deeply and carefully. This is especially true for multiple-choice questions that are preceded by a long prompt.

9. Subtle Negatives

One of the oldest tricks in the multiple-choice test writer's book is to subtly reverse the meaning of a question with a word like *not* or *except*. If you are not paying attention to each word in the question, you can easily be led astray by this trick. For instance, a common question format is, "Which of the following is…?" Obviously, if the question instead is, "Which of the following is not…?," then the answer will be quite different. Even worse, the test makers are aware of the potential for this mistake and will include one answer choice that would be correct if the question were not negated or reversed. A test taker who misses the reversal will find what he or she believes to be a correct answer and will be so confident that he or she will fail to reread the question and discover the original error. The only way to avoid this is to practice a wide variety of multiple-choice questions and to pay close attention to each and every word.

10. Reading Every Answer Choice

It may seem obvious, but you should always read every one of the answer choices! Too many test takers fall into the habit of scanning the question and assuming that they understand the question because they recognize a few key words. From there, they pick the first answer choice that answers the question they believe they have read. Test takers who read all of the answer choices might discover that one of the latter answer choices is actually *more* correct. Moreover, reading all of the answer choices can remind you of facts related to the question that can help you arrive at the correct answer. Sometimes, a misstatement or incorrect detail in one of the latter answer choices will trigger your memory of the subject and will enable you to find the right answer. Failing to read all of the answer choices is like not reading all of the items on a restaurant menu: you might miss out on the perfect choice.

11. Spot the Hedges

One of the keys to success on multiple-choice tests is paying close attention to every word. This is never truer than with words like almost, most, some, and sometimes. These words are called "hedges" because they indicate that a statement is not totally true or not true in every place and time. An absolute statement will contain no hedges, but in many subjects, the answers are not always straightforward or absolute. There are always exceptions to the rules in these subjects. For this reason, you should favor those multiple-choice questions that contain hedging language. The presence of qualifying words indicates that the author is taking special care with his or her words, which is certainly important when composing the right answer. After all, there are many ways to be wrong, but there is only one way to be right! For this reason, it is wise to avoid answers that are absolute when taking a multiple-choice test. An absolute answer is one that says things are either all one way or all another. They often include words like *every*, *always*, *best*, and *never*. If you are taking a multiple-choice test in a subject that doesn't lend itself to absolute answers, be on your guard if you see any of these words.

12. Long Answers

In many subject areas, the answers are not simple. As already mentioned, the right answer often requires hedges. Another common feature of the answers to a complex or subjective question are qualifying clauses, which are groups of words that subtly modify the meaning of the sentence. If the question or answer choice describes a rule to which there are exceptions or the subject matter is complicated, ambiguous, or confusing, the correct answer will require many words in order to be expressed clearly and accurately. In essence, you should not be deterred by answer choices that seem excessively long. Oftentimes, the author of the text will not be able to write the correct answer without

offering some qualifications and modifications. Your job is to read the answer choices thoroughly and completely and to select the one that most accurately and precisely answers the question.

13. Restating to Understand

Sometimes, a question on a multiple-choice test is difficult not because of what it asks but because of how it is written. If this is the case, restate the question or answer choice in different words. This process serves a couple of important purposes. First, it forces you to concentrate on the core of the question. In order to rephrase the question accurately, you have to understand it well. Rephrasing the question will concentrate your mind on the key words and ideas. Second, it will present the information to your mind in a fresh way. This process may trigger your memory and render some useful scrap of information picked up while studying.

14. True Statements

Sometimes an answer choice will be true in itself, but it does not answer the question. This is one of the main reasons why it is essential to read the question carefully and completely before proceeding to the answer choices. Too often, test takers skip ahead to the answer choices and look for true statements. Having found one of these, they are content to select it without reference to the question above. Obviously, this provides an easy way for test makers to play tricks. The savvy test taker will always read the entire question before turning to the answer choices. Then, having settled on a correct answer choice, he or she will refer to the original question and ensure that the selected answer is relevant. The mistake of choosing a correct-but-irrelevant answer choice is especially common on questions related to specific pieces of objective knowledge. A prepared test taker will have a wealth of factual knowledge at his or her disposal, and should not be careless in its application.

15. No Patterns

One of the more dangerous ideas that circulates about multiple-choice tests is that the correct answers tend to fall into patterns. These erroneous ideas range from a belief that B and C are the most common right answers, to the idea that an unprepared test-taker should answer "A-B-A-C-A-D-A-B-A." It cannot be emphasized enough that pattern-seeking of this type is exactly the WRONG way to approach a multiple-choice test. To begin with, it is highly unlikely that the test maker will plot the correct answers according to some predetermined pattern. The questions are scrambled and delivered in a random order. Furthermore, even if the test maker was following a pattern in the assignation of correct answers, there is no reason why the test taker would know which pattern he or she was using. Any attempt to discern a pattern in the answer choices is a waste of time and a distraction from the real work of taking the test. A test taker would be much better served by extra preparation before the test than by reliance on a pattern in the answers.

FREE DVD OFFER

Don't forget that doing well on your exam includes both understanding the test content and understanding how to use what you know to do well on the test. We offer a completely FREE Test Taking Tips DVD that covers world class test taking tips that you can use to be even more successful when you are taking your test.

All that we ask is that you email us your feedback about your study guide. To get your **FREE Test Taking Tips DVD**, email freedvd@studyguideteam.com with "FREE DVD" in the subject line and the following information in the body of the email:

- The title of your study guide.
- Your product rating on a scale of 1-5, with 5 being the highest rating.
- Your feedback about the study guide. What did you think of it?
- Your full name and shipping address to send your free DVD.

Introduction to the SHSAT

Function of the Test

The SHSAT is a standardized test that is used as the sole factor for admission to eight of New York City's Specialized High Schools. Fiorello H. LaGuardia High School is the only Specialized High School in New York City that does not require students to take the SHSAT exam as part of the admissions process. Students who are in the eighth or ninth grades who wish to attend one of these eight Specialized High Schools and who live in the five boroughs of New York City (Brooklyn, Manhattan, Queens, Staten Island, and The Bronx) must sit for this exam.

Test Administration

Each year, the SHSAT test is only offered in the month of October for eighth grade students and in the month of November for ninth grade students. Students who are interested in registering to take the exam can do so by talking with their school's guidance counselor. After students are registered, they will receive a test ticket to sit for the exam. Students who are sitting for the SHSAT exam must also rank (in order of priority) the Specialized High Schools that they would like to attend on their test ticket.

Students are able to take the SHSAT test twice—once in the eighth grade and once in the ninth grade—if they are not accepted to the Specialized High School of their choice after taking the exam in the eighth grade.

Students will be provided with the necessary accommodations for taking the exam, as long as the accommodations are permitted for the test. If necessary, mathematics glossaries can be provided in nine languages to students on the day of the exam.

Test Format

Students are given 180 minutes to complete the SHSAT, which is comprised of 57 questions in each of its two sections: English language arts (ELA) and math, as outlined in the table below. All of the reading and writing questions in the ELA section are multiple-choice and split between two categories. The first category requires students to utilize their revising and editing skills, while the second category assesses reading comprehension by asking students to extract information from various reading passages in order to answer associated questions. In the math section, there are 52 multiple-choice questions that deal with word and computational problems, as well as five grid-in questions that are *not* multiple-choice. These questions require students to provide correct numerical solutions to computational problems. Finally, all multiple-questions on both sections of the test have four answer choices per question, and both sections of the test each have ten unscored experimental questions that are used for field testing purposes for future iterations of the exam.

Sections of the SHSAT Test				
Subject Areas	**Questions (Multiple-Choice)***	**Question breakdown**		**Time Limit**
English Language Arts (ELA) Reading & Writing	57	9-11 revising/editing		180 minutes
		46-48 reading comprehension questions 3-4 informational passages 1-2 literary prose passages 1 poem		
		10 embedded field questions (unscored)		
Math	57	52 word & computational problems		
		5 grid-in questions*		
		10 embedded field questions (unscored)		
Total Questions:	114			

Students are required to provide correct numerical answers for these questions

Scoring

Individuals are not penalized for wrong answers or for questions that are left blank. After completing the test, each student is given a raw score that is based on the number of questions answered correctly. Those raw scores are then converted into three-digit composite scores (an 800 being the highest possible score). Scores are made available to the schools in March following the fall in which the exam was taken. For example, if a student takes the exam in the fall of 2017, his or her score will be released to the schools in the March of 2018.

Once the test results are in, all students who took the SHSAT exam are ranked in order by composite score from highest to lowest. Seats are then filled in each of the Specialized High Schools, in order, according to the students' first choices until all of the open seats for that academic year are filled. The number of available seats at each of the Specialized High Schools varies from year to year.

Recent/Future Developments

The SHSAT was revised in 2018 with changes mainly to the English Language arts (ELA) section. The amount of questions in the Revising/Editing section decreased from 20 to 9-11. The Reading Comprehension section previously included primarily informational texts to analyze. The update now includes informational passages as well as literary prose passages and one poem on which questions are based.

SHSAT Practice Test #1

Editing/Revising

Editing/Revising Part A

1. Read this sentence.

> Protestors filled the streets of the city. Because they were dissatisfied with the government's leadership.

How should this sentence be revised?

    a. Protestors filled the streets of the city, because they were dissatisfied with the government's leadership.
    b. Protesters, filled the streets of the city, because they were dissatisfied with the government's leadership.
    c. Because they were dissatisfied with the government's leadership protestors filled the streets of the city.
    d. Protestors filled the streets of the city because they were dissatisfied with the government's leadership.

2. Read this sentence.

> She's looking for a suitcase that can fit all of her clothes, shoes, accessory, and makeup.

How should the sentence be revised?

    a. Change *shoes* to **shoe**.
    b. Change the commas to semicolons.
    c. Change *makeup* to **makeups**.
    d. Change *accessory* to **accessories**.

3. Read this paragraph.

> (1) Early in my career, a master's teacher shared this thought with me: "Education is the last bastion of civility." (2) While I did not completely understand the scope of those words at the time, I have since come to realize the depth, breadth, truth, and significance of what he said. (3) Education provides society with a vehicle for raising it's children to be civil, decent, human beings with something valuable to contribute to the world. (4) It is really what makes us human and what distinguishes us as civilized creatures.

How should the paragraph be revised?

    a. Sentence 1: Move the period to outside the quotation marks.
    b. Sentence 2: Remove the comma after *time*.
    c. Sentence 3: Remove the apostrophe from *it's*.
    d. Sentence 4: Change *distinguishes* to **distinguished**.

4. Read this paragraph.

(1) George Washington Carver was an innovator, always thinking of new and better ways to do things and is most famous for his over three hundred uses for the peanut. (2) Toward the end of his career, Carver returns to his first love of art. (3) When Carver died, he left his money to help fund ongoing agricultural research. (4) Today, people still visit and study at the George Washington Carver Foundation at the Tuskegee Institute.

Which sentence contains an error in construction and should be revised?

    a. Sentence 1
    b. Sentence 2
    c. Sentence 3
    d. Sentence 4

Editing/Revising Part B

Read the text below and answer the questions following it.

(1) I have to admit that when my father bought a recreational vehicle (RV), I thought he was making a huge mistake. (2) I didn't really know anything about RVs, but I knew that my dad was as big a "city slicker" as there was. (3) In fact, I even thought he might have gone a little bit crazy. (4) On trips to the beach, he preferred to swim at the pool, and whenever he went hiking, he avoided touching any plants for fear that they might be poison ivy. (5) Why would this man, with an almost irrational fear of the outdoors, want a 40-foot camping behemoth?

(6) The RV was a great purchase for our family and brought us all closer together. (7) Every morning we would wake up, eat breakfast, and broke camp. (8) We laughed at our own comical attempts to back The Beast into spaces that seemed impossibly small. (9) We rejoiced as "hackers." (10) When things inevitably went wrong and we couldn't solve the problems on our own, we discovered the incredible helpfulness and friendliness of the RV community. (11) RV parks can be found in every US state. (12) We even made some new friends in the process.

(13) Above all, it allowed us to share adventures. (14) While travelling across America, which we could not have experienced in cars and hotels. (15) Enjoying a campfire on a chilly summer evening with the mountains of Glacier National Park in the background, or waking up early in the morning to see the sun rising over the distant spires of Arches National Park are memories that will always stay with me and our entire family.

5. Which of the following would be the best choice for sentence 3?

    a. Move the sentence to the beginning of the paragraph.
    b. Move the sentence so that it comes before the preceding sentence.
    c. Move the sentence to the end of the first paragraph.
    d. Omit the sentence.

6. Which transition phrase should be added to the beginning of sentence 6?

    a. Unfortunately,
    b. Not surprisingly,
    c. Furthermore,
    d. As it turns out,

7. Which is the best version of sentence 7?
   a. Every morning we woke up, eat breakfast, and broke camp.
   b. Every morning we would wake up, eat breakfast, and break camp.
   c. Every morning would we wake up, eat breakfast, and break camp?
   d. Every morning we are waking up, eating breakfast, and breaking camp.

8. Which revision includes the most precise version of sentence 9?
   a. "Hackers" rejoiced when they saw us.
   b. To a nagging problem of technology, we rejoiced as "hackers."
   c. We rejoiced when we figured out how to "hack" a solution to a nagging technological problem.
   d. To "hack" our way to a solution, we had to rejoice.

9. Which sentence presents information not relevant to the main topic of the second paragraph and should be removed?
   a. Sentence 9
   b. Sentence 10
   c. Sentence 11
   d. Sentence 12

10. Which is the best way to combine sentences 13 and 14?
    a. Above all and while traveling across America, it allowed us to share adventures which we could not have experienced in cars and hotels.
    b. Above all, it allowed us to share adventures while traveling across America, which we could not have experienced in cars and hotels.
    c. Above all, it allowed us to share adventures; while traveling across America, which we could not have experienced in cars and hotels.
    d. Above all, it allowed us to share adventures—while traveling across America, which we could not have experienced in cars and hotels.

11. Which sentence should be added after sentence 15 to conclude the passage?
    a. My dad now loves the outdoors.
    b. My family bonded over the mistakes we made while traveling in our RV.
    c. Those are also memories that my siblings and I have now shared with our children.
    d. My siblings and I never stay in hotels now that we are grown.

Reading Comprehension

Read the following poem and answer questions 1–6.

Two roads diverged in a yellow wood,
And sorry I could not travel both
And be one traveler, long I stood
And looked down one as far as I could
To where it bent in the undergrowth;   5
Then took the other, as just as fair,
And having perhaps the better claim,
Because it was grassy and wanted wear;
Though as for that the passing there
Had worn them really about the same,   10
And both that morning equally lay
In leaves no step had trodden black.
Oh, I kept the first for another day!
Yet knowing how way leads on to way,
I doubted if I should ever come back.   15
I shall be telling this with a sigh
Somewhere ages and ages hence:
Two roads diverged in a wood, and I—
I took the one less traveled by,
And that has made all the difference.   20

Robert Frost, "The Road Not Taken"

1. Which option best expresses the symbolic meaning of the "road" and the overall theme?
    a. A divergent spot where the traveler had to choose the correct path to his destination
    b. A choice between good and evil that the traveler needs to make
    c. The traveler's struggle between his lost love and his future prospects
    d. Life's journey and the choices with which humans are faced

2. Which line best contributes to the idea that the second path was *less traveled*?
    a. 5
    b. 4
    c. 8
    d. 10

3. What could the author mean by the phrase *knowing how way leads on to way*?
    a. Once a path is chosen, other choices flow from that
    b. Choosing the second path could lead to getting lost
    c. The second path could lead back to the first path
    d. The two different paths could reach the same end

4. How many travelers are there in the poem?
    a. 1
    b. 2
    c. 3
    d. 4

5. What is the time of day in the poem?
    a. Night
    b. Morning
    c. Evening
    d. Afternoon

6. Which road did the traveler take?
    a. The first
    b. The second
    c. Both
    d. Neither

Questions 7–12 are based upon the following passage:

This excerpt is adaptation of Robert Louis Stevenson's The Strange Case of Dr. Jekyll and Mr. Hyde.

"Did you ever come across a protégé of his—one Hyde?" He asked.

"Hyde?" repeated Lanyon. "No. Never heard of him. Since my time."

That was the amount of information that the lawyer carried back with him to the great, dark bed on which he tossed to and fro until the small hours of the morning began to grow large. It was a night of little ease to his toiling mind, toiling in mere darkness and besieged by questions.

Six o'clock struck on the bells of the church that was so conveniently near to Mr. Utterson's dwelling, and still he was digging at the problem. Hitherto it had touched him on the intellectual side alone; but; but now his imagination also was engaged, or rather enslaved; and as he lay and tossed in the gross darkness of the night in the curtained room, Mr. Enfield's tale went by before his mind in a scroll of lighted pictures. He would be aware of the great field of lamps in a nocturnal city; then of the figure of a man walking swiftly; then of a child running from the doctor's; and then these met, and that human Juggernaut trod the child down and passed on regardless of her screams. Or else he would see a room in a rich house, where his friend lay asleep, dreaming and smiling at his dreams; and then the door of that room would be opened, the curtains of the bed plucked apart, the sleeper recalled, and, lo! There would stand by his side a figure to whom power was given, and even at that dead hour he must rise and do its bidding. The figure in these two phrases haunted the lawyer all night; and if at anytime he dozed over, it was but to see it glide more stealthily through sleeping houses, or move the more swiftly, and still the more smoothly, even to dizziness, through wider labyrinths of lamplighted city, and at every street corner crush a child and leave her screaming. And still the figure had no face by which he might know it; even in his dreams it had no face, or one that baffled him and melted before his eyes; and thus there it was that there sprung up and grew apace in the lawyer's mind a singularly strong, almost an inordinate, curiosity to behold the features of the real Mr. Hyde. If he could but once set eyes on him, he thought the mystery would lighten and perhaps roll altogether away, as was the habit of mysterious things when well examined. He might see a reason for his friend's strange preference or bondage, and even for the startling clauses of the will. And at least it would be a face worth seeing: the face of a man who was without bowels of mercy: a face which had but to show itself to raise up, in the mind of the unimpressionable Enfield, a spirit of enduring hatred.

From that time forward, Mr. Utterson began to haunt the door in the by street of shops. In the morning before office hours, at noon when business was plenty of time scarce, at night under the face of the full city moon, by all lights and at all hours of solitude or concourse, the lawyer was to be found on his chosen post.

"If he be Mr. Hyde," he had thought, "I should be Mr. Seek."

7. What is the purpose of the use of repetition in the following passage?
   It was a night of little ease to his toiling mind, toiling in mere darkness and besieged by questions.

   a. It serves as a demonstration of the mental state of Mr. Lanyon.
   b. It is reminiscent of the church bells that are mentioned in the story.
   c. It mimics Mr. Utterson's ambivalence.
   d. It emphasizes Mr. Utterson's anguish in failing to identify Hyde's whereabouts.

8. What is the setting of the story in this passage?
   a. In the city
   b. On the countryside
   c. In a jail
   d. In a mental health facility

9. What can one infer about the meaning of the word "Juggernaut" from the author's use of it in the passage?
   a. It is an apparition that appears at daybreak.
   b. It scares children.
   c. It is associated with space travel.
   d. Mr. Utterson finds it soothing.

10. What is the definition of the word *haunt* in the following passage?
    From that time forward, Mr. Utterson began to haunt the door in the by street of shops. In the morning before office hours, at noon when business was plenty of time scarce, at night under the face of the full city moon, by all lights and at all hours of solitude or concourse, the lawyer was to be found on his chosen post.

    a. To levitate
    b. To constantly visit
    c. To terrorize
    d. To daunt

11. The phrase *labyrinths of lamplighted city* contains an example of what?
    a. Hyperbole
    b. Simile
    c. Metaphor
    d. Alliteration

12. What can one reasonably conclude from the final comment of this passage?
    "If he be Mr. Hyde," he had thought, "I should be Mr. Seek."

    a. The speaker is considering a name change.
    b. The speaker is experiencing an identity crisis.
    c. The speaker has mistakenly been looking for the wrong person.
    d. The speaker intends to continue to look for Hyde.

Questions 13–18 are based upon the following passage:

This excerpt is an adaptation of Jonathan Swift's Gulliver's Travels into Several Remote Nations of the World.

My gentleness and good behaviour had gained so far on the emperor and his court, and indeed upon the army and people in general, that I began to conceive hopes of getting my liberty in a short time. I took all possible methods to cultivate this favourable disposition. The natives came, by degrees, to be less apprehensive of any danger from me. I would sometimes lie down, and let five or six of them dance on my hand; and at last the boys and girls would venture to come and play at hide-and-seek in my hair. I had now made a good progress in understanding and speaking the language. The emperor had a mind one day to entertain me with several of the country shows, wherein they exceed all nations I have known, both for dexterity and magnificence. I was diverted with none so much as that of the rope-dancers, performed upon a slender white thread, extended about two feet, and twelve inches from the ground. Upon which I shall desire liberty, with the reader's patience, to enlarge a little.

This diversion is only practised by those persons who are candidates for great employments, and high favour at court. They are trained in this art from their youth, and are not always of noble birth, or liberal education. When a great office is vacant, either by death or disgrace (which often happens,) five or six of those candidates petition the emperor to entertain his majesty and the court with a dance on the rope; and whoever jumps the highest, without falling, succeeds in the office. Very often the chief ministers themselves are commanded to show their skill, and to convince the emperor that they have not lost their faculty. Flimnap, the treasurer, is allowed to cut a caper on the straight rope, at least an inch higher than any other lord in the whole empire. I have seen him do the summerset several times together, upon a trencher fixed on a rope which is no thicker than a common packthread In England. My friend Reldresal, principal secretary for private affairs, is, in my opinion, if I am not partial, the second after the treasurer; the rest of the great officers are much upon a par.

13. Which of the following statements best summarize the central purpose of this text?
    a. Gulliver details his fondness for the archaic yet interesting practices of his captors.
    b. Gulliver conjectures about the intentions of the aristocratic sector of society.
    c. Gulliver becomes acquainted with the people and practices of his new surroundings.
    d. Gulliver's differences cause him to become penitent around new acquaintances.

14. What is the word *principal* referring to in the following text?
    My friend Reldresal, principal secretary for private affairs, is, in my opinion, if I am not partial, the second after the treasurer; the rest of the great officers are much upon a par.

    a. Primary or chief
    b. An acolyte
    c. An individual who provides nurturing
    d. One in a subordinate position

15. What can the reader infer from this passage?
    I would sometimes lie down, and let five or six of them dance on my hand; and at last the boys and girls would venture to come and play at hide-and-seek in my hair.

    a. The children tortured Gulliver.
    b. Gulliver traveled because he wanted to meet new people.
    c. Gulliver is considerably larger than the children who are playing around him.
    d. Gulliver has a genuine love and enthusiasm for people of all sizes.

16. What is the significance of the word *mind* in the following passage?
    The emperor had a mind one day to entertain me with several of the country shows, wherein they exceed all nations I have known, both for dexterity and magnificence.

    a. The ability to think
    b. A collective vote
    c. A definitive decision
    d. A mythological question

17. Which of the following assertions does not support the fact that games are a commonplace event in this culture?
    a. My gentlest and good behavior . . . short time.
    b. They are trained in this art from their youth . . . liberal education.
    c. Very often the chief ministers themselves are commanded to show their skill . . . not lost their faculty.
    d. Flimnap, the treasurer, is allowed to cut a caper on the straight rope . . . higher than any other lord in the whole empire.

18. How do the roles of Flimnap and Reldresal serve as evidence of the community's emphasis in regards to the correlation between physical strength and leadership abilities?
    a. Only children used Gulliver's hands as a playground.
    b. The two men who exhibited superior abilities held prominent positions in the community.
    c. Only common townspeople, not leaders, walk the straight rope.
    d. No one could jump higher than Gulliver.

Questions 19–24 are based upon the following passage:

This excerpt is adaptation from Abraham Lincoln's Address Delivered at the Dedication of the Cemetery at Gettysburg, November 19, 1863.

Four score and seven years ago our fathers brought forth on this continent, a new nation, conceived in liberty, and dedicated to the proposition that all men are created equal.

Now we are engaged in a great civil war, testing whether that nation, or any nation so conceived and so dedicated, can long endure. We are met on a great battlefield of that war. We have come to dedicate a portion of that field, as a final resting place for those who here gave their lives that this nation might live. It is altogether fitting and proper that we should do this.

But, in a larger sense, we cannot dedicate—we cannot consecrate that we cannot hallow—this ground. The brave men, living and dead, who struggled here, have consecrated it, far above our poor power to add or detract. The world will little note, nor long remember what we say here, but it can never forget what they did here. It is for us the living, rather, to be dedicated here to the unfinished work which they who fought here have thus far so nobly advanced. It is rather for us to be here and dedicated to the great task remaining before us—that from these honored dead we take increased devotion to that cause for which they gave the last full measure of devotion—that we here highly resolve that these dead shall not have died in vain—that these this nation, under God, shall have a new birth of freedom—and that government of people, by the people, for the people, shall not perish from the earth.

19. The best description for the phrase *four score and seven years ago* is which of the following?
    a. A unit of measurement
    b. A period of time
    c. A literary movement
    d. A statement of political reform

20. What is the setting of this text?
    a. A battleship off of the coast of France
    b. A desert plain on the Sahara Desert
    c. A battlefield in North America
    d. The residence of Abraham Lincoln

21. Which war is Abraham Lincoln referring to in the following passage?
    Now we are engaged in a great civil war, testing whether that nation, or any nation so conceived and so dedicated, can long endure.

    a. World War I
    b. The War of the Spanish Succession
    c. World War II
    d. The American Civil War

22. What message is the author trying to convey through this address?
a. The audience should consider the death of the people that fought in the war as an example and perpetuate the ideals of freedom that the soldiers died fighting for.
b. The audience should honor the dead by establishing an annual memorial service.
c. The audience should form a militia that would overturn the current political structure.
d. The audience should forget the lives that were lost and discredit the soldiers.

23. Which rhetorical device is being used in the following passage?

...we here highly resolve that these dead shall not have died in vain—that these this nation, under God, shall have a new birth of freedom—and that government of people, by the people, for the people, shall not perish from the earth.

a. Antimetabole
b. Antiphrasis
c. Anaphora
d. Epiphora

24. What is the effect of Lincoln's statement in the following passage?

But, in a larger sense, we cannot dedicate—we cannot consecrate that we cannot hallow—this ground. The brave men, living and dead, who struggled here, have consecrated it, far above our poor power to add or detract.

a. His comparison emphasizes the great sacrifice of the soldiers who fought in the war.
b. His comparison serves as a remainder of the inadequacies of his audience.
c. His comparison serves as a catalyst for guilt and shame among audience members.
d. His comparison attempts to illuminate the great differences between soldiers and civilians.

Questions 25–29 are based on the following passage:

"MANKIND being originally equals in the order of creation, the equality could only be destroyed by some subsequent circumstance; the distinctions of rich, and poor, may in a great measure be accounted for, and that without having recourse to the harsh ill sounding names of oppression and avarice. Oppression is often the consequence, but seldom or never the means of riches; and though avarice will preserve a man from being necessitously poor, it generally makes him too timorous to be wealthy.

But there is another and greater distinction for which no truly natural or religious reason can be assigned, and that is, the distinction of men into KINGS and SUBJECTS. Male and female are the distinctions of nature, good and bad the distinctions of heaven; but how a race of men came into the world so exalted above the rest, and distinguished like some new species, is worth enquiring into, and whether they are the means of happiness or of misery to mankind.

In the early ages of the world, according to the scripture chronology, there were no kings; the consequence of which was there were no wars; it is the pride of kings which throw mankind into confusion Holland without a king hath enjoyed more peace for this last century than any of the monarchical governments in Europe. Antiquity favors the same remark; for the quiet and rural lives of the first patriarchs hath a happy something in them, which vanishes away when we come to the history of Jewish royalty.

Government by kings was first introduced into the world by the Heathens, from whom the children of Israel copied the custom. It was the most prosperous invention the Devil ever set on foot for the promotion of idolatry. The Heathens paid divine honors to their deceased kings, and the Christian world hath improved on the plan by doing the same to their living ones. How impious is the title of sacred majesty applied to a worm, who in the midst of his splendor is crumbling into dust!

As the exalting one man so greatly above the rest cannot be justified on the equal rights of nature, so neither can it be defended on the authority of scripture; for the will of the Almighty, as declared by Gideon and the prophet Samuel, expressly disapproves of government by kings. All anti-monarchical parts of scripture have been very smoothly glossed over in monarchical governments, but they undoubtedly merit the attention of countries, which have their governments yet to form. "Render unto Caesar the things which are Caesar's" is the scripture doctrine of courts, yet it is no support of monarchical government, for the Jews at that time were without a king, and in a state of vassalage to the Romans.

Near three thousand years passed away from the Mosaic account of the creation, till the Jews under a national delusion requested a king. Till then their form of government (except in extraordinary cases, where the Almighty interposed) was a kind of republic administered by a judge and the elders of the tribes. Kings they had none, and it was held sinful to acknowledge any being under that title but the Lord of Hosts. And when a man seriously reflects on the idolatrous homage which is paid to the persons of Kings, he need not wonder, that the Almighty ever jealous of his honor, should disapprove of a form of government which so impiously invades the prerogative of heaven.

Excerpt From: Thomas Paine. "Common Sense."

25. According to passage, what role does avarice, or greed, play in poverty?
    a. It can make a man very wealthy
    b. It is the consequence of wealth
    c. Avarice can prevent a man from being poor, but too fearful to be very wealthy
    d. Avarice is what drives a person to be very wealthy

26. Of these distinctions, which does the author believe to be beyond natural or religious reason?
    a. Good and bad
    b. Male and female
    c. Human and animal
    d. King and subjects

27. According to the passage, what are the Heathens responsible for?
    a. Government by kings
    b. Quiet and rural lives of patriarchs
    c. Paying divine honors to their living kings
    d. Equal rights of nature

28. Which of the following best states Paine's rationale for the denouncement of monarchy?
    a. It is against the laws of nature
    b. It is against the equal rights of nature and is denounced in scripture
    c. Despite scripture, a monarchal government is unlawful
    d. Neither the law nor scripture denounce monarchy

29. Based on the passage, what is the best definition of the word *idolatrous*?
    a. Worshipping heroes
    b. Being deceitful
    c. Sinfulness
    d. Engaging in illegal activities

Questions 30–35 are based upon the following passage:

This excerpt is adapted from "What to the Slave is the Fourth of July?" Rochester, New York July 5, 1852.

Fellow citizens—Pardon me, and allow me to ask, why am I called upon to speak here today? What have I, or those I represent, to do with your national independence? Are the great principles of political freedom and of natural justice embodied in that Declaration of Independence, Independence extended to us? And am I therefore called upon to bring our humble offering to the national altar, and to confess the benefits, and express devout gratitude for the blessings, resulting from your independence to us?

Would to God, both for your sakes and ours, ours that an affirmative answer could be truthfully returned to these questions! Then would my task be light, and my burden easy and delightful. For who is there so cold that a nation's sympathy could not warm him? Who so obdurate and dead to the claims of gratitude that would not thankfully acknowledge such priceless benefits? Who so stolid and selfish, that would not give his voice to swell the hallelujahs of a nation's jubilee, when the chains of servitude had been torn from his limbs? I am not that man. In a case like that, the dumb my eloquently speak, and the lame man leap as an hart.

But, such is not the state of the case. I say it with a sad sense of the disparity between us. I am not included within the pale of this glorious and anniversary. Oh pity! Your high independence only reveals the immeasurable distance between us. The blessings in which you this day rejoice, I do not enjoy in common. The rich inheritance of justice, liberty, prosperity, and independence, bequeathed by your fathers, is shared by *you*, not by *me*. This Fourth of July is *yours*, not *mine*. You may rejoice, *I* must mourn. To drag a man in fetters into the grand illuminated temple of liberty, and call upon him to join you in joyous anthems, were inhuman mockery and sacrilegious irony. Do you mean, citizens, to mock me, by asking me to speak today? If so there is a parallel to your conduct. And let me warn you that it is dangerous to copy the example of a nation whose crimes, towering up to heaven, were thrown down by the breath of the Almighty, burying that nation and irrecoverable ruin! I can today take up the plaintive lament of a peeled and woe-smitten people.

By the rivers of Babylon, there we sat down. Yea! We wept when we remembered Zion. We hanged our harps upon the willows in the midst thereof. For there, they that carried us away captive, required of us a song; and they who wasted us required of us mirth,

saying, "Sing us one of the songs of Zion." How can we sing the Lord's song in a strange land? If I forget thee, O Jerusalem, let my right hand forget her cunning. If I do not remember thee, let my tongue cleave to the roof of my mouth.

30. What is the tone of the first paragraph of this passage?
   a. Exasperated
   b. Inclusive
   c. Contemplative
   d. Nonchalant

31. Which word CANNOT be used synonymously with the term *obdurate* as it is conveyed in the text below?

   Who so obdurate and dead to the claims of gratitude, that would not thankfully acknowledge such priceless benefits?

   a. Steadfast
   b. Stubborn
   c. Contented
   d. Unwavering

32. What is the central purpose of this text?
   a. To demonstrate the author's extensive knowledge of the Bible
   b. To address the feelings of exclusion expressed by African Americans after the establishment of the Fourth of July holiday
   c. To convince wealthy landowners to adopt new holiday rituals
   d. To explain why minorities often relished the notion of segregation in government institutions

33. Which statement serves as evidence for the question above?
   a. By the rivers of Babylon...down.
   b. Fellow citizens...today.
   c. I can...woe-smitten people.
   d. The rich inheritance of justice...*not by me.*

34. The statement below features an example of which of the following literary devices?
   Oh pity! Your high independence only reveals the immeasurable distance between us.

   a. Assonance
   b. Parallelism
   c. Amplification
   d. Hyperbole

35. The speaker's use of biblical references, such as "rivers of Babylon" and the "songs of Zion," helps the reader to do all EXCEPT which of the following?
   a. Identify with the speaker using common text
   b. Convince the audience that injustices have been committed by referencing another group of people who have been previously affected by slavery
   c. Display the equivocation of the speaker and those that he represents
   d. Appeal to the listener's sense of humanity

Questions 36–41 are based upon the following passage:

This excerpt is adaptation from *Our Vanishing Wildlife,* by William T. Hornaday

Three years ago, I think there were not many bird-lovers in the United States, who believed it possible to prevent the total extinction of both egrets from our fauna. All the known rookeries accessible to plume-hunters had been totally destroyed. Two years ago, the secret discovery of several small, hidden colonies prompted William Dutcher, President of the National Association of Audubon Societies, and Mr. T. Gilbert Pearson, Secretary, to attempt the protection of those colonies. With a fund contributed for the purpose, wardens were hired and duly commissioned. As previously stated, one of those wardens was shot dead in cold blood by a plume hunter. The task of guarding swamp rookeries from the attacks of money-hungry desperadoes to whom the accursed plumes were worth their weight in gold, is a very chancy proceeding. There is now one warden in Florida who says that "before they get my rookery they will first have to get me."

Thus far the protective work of the Audubon Association has been successful. Now there are twenty colonies, which contain all told, about 5,000 egrets and about 120,000 herons and ibises which are guarded by the Audubon wardens. One of the most important is on Bird Island, a mile out in Orange Lake, central Florida, and it is ably defended by Oscar E. Baynard. To-day, the plume hunters who do not dare to raid the guarded rookeries are trying to study out the lines of flight of the birds, to and from their feeding-grounds, and shoot them in transit. Their motto is—"Anything to beat the law, and get the plumes." It is there that the state of Florida should take part in the war.

The success of this campaign is attested by the fact that last year a number of egrets were seen in eastern Massachusetts—for the first time in many years. And so to-day the question is, can the wardens continue to hold the plume-hunters at bay?

36. The author's use of first person pronoun in the following text does NOT have which of the following effects?

Three years ago, I think there were not many bird-lovers in the United States, who believed it possible to prevent the total extinction of both egrets from our fauna.

a. The phrase *I think* acts as a sort of hedging, where the author's tone is less direct and/or absolute.
b. It allows the reader to more easily connect with the author.
c. It encourages the reader to empathize with the egrets.
d. It distances the reader from the text by overemphasizing the story.

37. What purpose does the quote serve at the end of the first paragraph?
a. The quote shows proof of a hunter threatening one of the wardens.
b. The quote lightens the mood by illustrating the colloquial language of the region.
c. The quote provides an example of a warden protecting one of the colonies.
d. The quote provides much needed comic relief in the form of a joke.

38. What is the meaning of the word *rookeries* in the following text?

  > To-day, the plume hunters who do not dare to raid the guarded rookeries are trying to study out the lines of flight of the birds, to and from their feeding-grounds, and shoot them in transit.

  a. Houses in a slum area
  b. A place where hunters gather to trade tools
  c. A place where wardens go to trade stories
  d. A colony of breeding birds

39. What is on Bird Island?
  a. Hunters selling plumes
  b. An important bird colony
  c. Bird Island Battle between the hunters and the wardens
  d. An important egret with unique plumes

40. What is the main purpose of the passage?
  a. To persuade the audience to act in preservation of the bird colonies
  b. To show the effect hunting egrets has had on the environment
  c. To argue that the preservation of bird colonies has had a negative impact on the environment.
  d. To demonstrate the success of the protective work of the Audubon Association

41. Why are hunters trying to study the lines of flight of the birds?
  a. To study ornithology, one must know the lines of flight that birds take.
  b. To help wardens preserve the lives of the birds
  c. To have a better opportunity to hunt the birds
  d. To builds their homes under the lines of flight because they believe it brings good luck

Questions 42–46 are based upon the following passage:

This excerpt is an adaptation from The Life-Story of Insects, *by Geo H. Carpenter.*

  Insects as a whole are preeminently creatures of the land and the air. This is shown not only by the possession of wings by a vast majority of the class, but by the mode of breathing to which reference has already been made, a system of branching air-tubes carrying atmospheric air with its combustion-supporting oxygen to all the insect's tissues. The air gains access to these tubes through a number of paired air-holes or spiracles, arranged segmentally in series.

  It is of great interest to find that, nevertheless, a number of insects spend much of their time under water. This is true of not a few in the perfect winged state, as for example aquatic beetles and water-bugs ('boatmen' and 'scorpions') which have some way of protecting their spiracles when submerged, and, possessing usually the power of flight, can pass on occasion from pond or stream to upper air. But it is advisable in connection with our present subject to dwell especially on some insects that remain continually under water till they are ready to undergo their final moult and attain the winged state, which they pass entirely in the air. The preparatory instars of such insects are aquatic; the adult instar is aerial. All may-flies, dragon-flies, and caddis-flies, many beetles and two-winged flies, and a few moths thus divide their life-story between the water and the air. For the present we confine attention to the Stone-flies, the May-flies, and the

Dragon-flies, three well-known orders of insects respectively called by systematists the Plecoptera, the Ephemeroptera and the Odonata.

In the case of many insects that have aquatic larvae, the latter are provided with some arrangement for enabling them to reach atmospheric air through the surface-film of the water. But the larva of a stone-fly, a dragon-fly, or a may-fly is adapted more completely than these for aquatic life; it can, by means of gills of some kind, breathe the air dissolved in water.

42. Which statement best details the central idea in this passage?
   a. It introduces certain insects that transition from water to air.
   b. It delves into entomology, especially where gills are concerned.
   c. It defines what constitutes as insects' breathing.
   d. It invites readers to have a hand in the preservation of insects.

43. Which definition most closely relates to the usage of the word *moult* in the passage?
   a. An adventure of sorts, especially underwater
   b. Mating act between two insects
   c. The act of shedding part or all of the outer shell
   d. Death of an organism that ends in a revival of life

44. What is the purpose of the first paragraph in relation to the second paragraph?
   a. The first paragraph serves as a cause and the second paragraph serves as an effect.
   b. The first paragraph serves as a contrast to the second.
   c. The first paragraph is a description for the argument in the second paragraph.
   d. The first and second paragraphs are merely presented in a sequence.

45. What does the following sentence most nearly mean?
   The preparatory instars of such insects are aquatic; the adult instar is aerial.

   a. The volume of water is necessary to prep the insect for transition rather than the volume of the air.
   b. The abdomen of the insect is designed like a star in the water as well as the air.
   c. The stage of preparation in between molting is acted out in the water, while the last stage is in the air.
   d. These insects breathe first in the water through gills, yet continue to use the same organs to breathe in the air.

46. Which of the statements reflect information that one could reasonably infer based on the author's tone?
   a. The author's tone is persuasive and attempts to call the audience to action.
   b. The author's tone is passionate due to excitement over the subject and personal narrative.
   c. The author's tone is informative and exhibits interest in the subject of the study.
   d. The author's tone is somber, depicting some anger at the state of insect larvae.

Math

1. If $6t + 4 = 16$, what is t?
    a. 1
    b. 2
    c. 3
    d. 4

2. The variable y is directly proportional to x. If $y = 3$ when $x = 5$, then what is y when $x = 20$?
    a. 10
    b. 12
    c. 14
    d. 16

3. A line passes through the point (1, 2) and crosses the y-axis at $y = 1$. Which of the following is an equation for this line?
    a. $y = 2x$
    b. $y = x + 1$
    c. $x + y = 1$
    d. $y = \frac{x}{2} - 2$

4. There are $4x + 1$ treats in each party favor bag. If a total of $60x + 15$ treats are distributed, how many bags are given out?
    a. 15
    b. 16
    c. 20
    d. 22

5. An accounting firm charted its income on the following pie graph. If the total income for the year was $500,000, how much of the income was received from Audit and Taxation Services?

Income

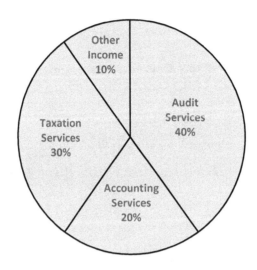

a. $200,000
b. $350,000
c. $150,000
d. $300,000

6. What are the roots of $x^2 + x - 2$?
   a. 1 and -2
   b. -1 and 2
   c. 2 and -2
   d. 9 and 13

7. What is the y-intercept of $y = x^{5/3} + (x - 3)(x + 1)$?
   a. 3.5
   b. 7.6
   c. -3
   d. -15.1

8. The phone bill is calculated each month using the equation $c = 50g + 75$. The cost of the phone bill per month is represented by c, and g represents the gigabytes of data used that month. What is the value and interpretation of the slope of this equation?
   a. 75 dollars per day
   b. 75 gigabytes per day
   c. 50 dollars per day
   d. 50 dollars per gigabyte

9. $(4x^2y^4)^{\frac{3}{2}}$ can be simplified to which of the following?

a. $8x^3y^6$

b. $4x^{\frac{5}{2}}y$

c. $4xy$

d. $32x^{\frac{7}{2}}y^{\frac{11}{2}}$

10. If $\sqrt{1+x} = 4$, what is x?

a. 10
b. 15
c. 20
d. 25

11. Suppose $\frac{x+2}{x} = 2$. What is x?

a. -1
b. 0
c. 2
d. 4

12. Which graph will be a line parallel to the graph of $y = 3x - 2$?

a. $2y - 6x = 2$
b. $y - 4x = 4$
c. $3y = x - 2$
d. $2x - 2y = 2$

13. A rectangle has a length that is 5 feet longer than three times its width. If the perimeter is 90 feet, what is the length in feet?

a. 10
b. 20
c. 25
d. 35

14. Five students take a test. The scores of the first four students are 80, 85, 75, and 60. If the median score is 80, which of the following could NOT be the score of the fifth student?

a. 60
b. 80
c. 85
d. 100

15. In an office, there are 50 workers. A total of 60% of the workers are women, and the chances of a woman wearing a skirt is 50%. If no men wear skirts, how many workers are wearing skirts?

a. 12
b. 15
c. 16
d. 20

16. Ten students take a test. Five students get a 50. Four students get a 70. If the average score is 55, what was the last student's score?

    a. 20
    b. 40
    c. 50
    d. 60

17. A company invests $50,000 in a building where they can produce saws. If the cost of producing one saw is $40, then which function expresses the amount of money the company pays? The variable y is the money paid and x is the number of saws produced.

    a. $y = 50{,}000x + 40$
    b. $y + 40 = x - 50{,}000$
    c. $y = 40x - 50{,}000$
    d. $y = 40x + 50{,}000$

18. A six-sided die is rolled. What is the probability that the roll is 1 or 2?

    a. $\frac{1}{6}$

    b. $\frac{1}{4}$

    c. $\frac{1}{3}$

    d. $\frac{1}{2}$

19. A line passes through the origin and through the point (-3, 4). What is the slope of the line?

    a. $-\frac{4}{3}$

    b. $-\frac{3}{4}$

    c. $\frac{4}{3}$

    d. $\frac{3}{4}$

20. An equilateral triangle has a perimeter of 18 feet. If a square whose sides have the same length as one side of the triangle is built, what will be the area of the square?

    a. 6 square feet
    b. 36 square feet
    c. 256 square feet
    d. 1000 square feet

21. Change $3\frac{3}{5}$ to a decimal.

    a. 3.6
    b. 4.67
    c. 5.3
    d. 0.28

22. The chart below shows the average car sales for the months of July through December for two different car dealers. What is the average number of cars sold in the given time period for Dealer 1?

a. 7
b. 11
c. 9
d. 8

23. The soccer team is selling donuts to raise money to buy new uniforms. For every box of donuts they sell, the team receives $3 towards their new uniforms. There are 15 people on the team. How many boxes does each player need to sell in order to raise $270 for their new uniforms?

a. 6
b. 30
c. 18
d. 5

24. What is the value of x in the diagram below?

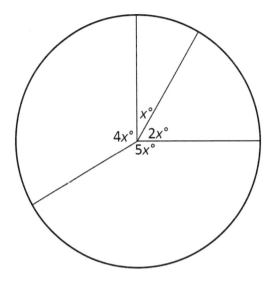

    a. 60
    b. 50
    c. 30
    d. 36

25. What is the probability of randomly picking the winner and runner-up from a race of 4 horses and distinguishing which is the winner?

    a. $\frac{1}{4}$

    b. $\frac{1}{2}$

    c. $\frac{1}{16}$

    d. $\frac{1}{12}$

26. A figure skater is facing north when she begins to spin to her right. She spins 2250 degrees. Which direction is she facing when she finishes her spin?

    a. North
    b. South
    c. East
    d. West

27. What is the measure of angle P?

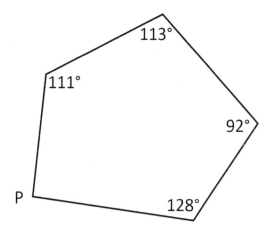

    a. 84 degrees
    b. 92 degrees
    c. 96 degrees
    d. 113 degrees

28. The expression $\frac{x-4}{x^2-6x+8}$ is undefined for what value(s) of x?
    a. 4 and 2
    b. -4 and -2
    c. 2
    d. 4

29. Nina has a jar where she puts her loose change at the end of each day. There are 13 quarters, 25 dimes, 18 nickels, and 30 pennies in the jar. If she chooses a coin at random, what is the probability that the coin will not be a penny or a dime?
    a. 0.36
    b. 0.64
    c. 0.56
    d. 0.34

30. The Cross family is planning a trip to Florida. They will be taking two cars for the trip. One car gets 18 miles to the gallon of gas. The other car gets 25 miles to the gallon. If the total trip to Florida is 450 miles, and the cost of gas is $2.49/gallon, how much will the gas cost for both cars to complete the trip?
    a. $43.00
    b. $44.82
    c. $107.07
    d. $32.33

31. Add and express in reduced form $\frac{14}{33} + \frac{10}{11}$.

   a. $\frac{2}{11}$

   b. $\frac{6}{11}$

   c. $\frac{4}{3}$

   d. $\frac{44}{33}$

32. 32 is 25% of what number?
   a. 64
   b. 128
   c. 12.65
   d. 8

33. If $3x - 4 + 5x = 8 - 10x$, what is the value of x?
   a. 6
   b. -6
   c. 0.5
   d. 0.67

34. Given the triangle below, find the value of x if $y = 21$.

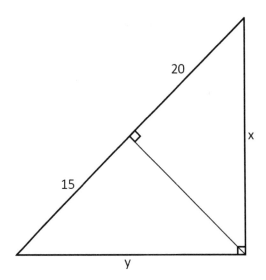

   a. 35
   b. 28
   c. 25
   d. 26

35. Of the given sets of coordinates below, which one lies on the line that is perpendicular to $y = 2x - 3$ and passes through the point $(0, 5)$?
 a. $(2, 4)$
 b. $(-2, 7)$
 c. $(4, -3)$
 d. $(-6, 10)$

36. Which of the following is a factor of both $x^2 + 4x + 4$ and $x^2 - x - 6$?
 a. $x - 3$
 b. $x + 2$
 c. $x - 2$
 d. $x + 3$

37. If $g(x) = x^3 - 3x^2 - 2x + 6$ and $f(x) = 2$, then what is $g(f(x))$?
 a. -26
 b. 6
 c. $2x^3 - 6x^2 - 4x + 12$
 d. -2

38. Which of the following is an equivalent measurement for 1.3 cm?
 a. 0.13 m
 b. 0.013 m
 c. 0.13 mm
 d. 0.013 mm

39. Divide $1,015 \div 1.4$.
 a. 7,250
 b. 725
 c. 7.25
 d. 72.50

40. What is the solution to the following system of equations?
$$x^2 - 2x + y = 8$$
$$x - y = -2$$
 a. $(-2, 3)$
 b. There is no solution.
 c. $(-2, 0) \ (1, 3)$
 d. $(-2, 0) \ (3, 5)$

41. Which of the following shows the correct result of simplifying the following expression:
$$(7n + 3n^3 + 3) + (8n + 5n^3 + 2n^4)$$
 a. $9n^4 + 15n - 2$
 b. $2n^4 + 5n^3 + 15n - 2$
 c. $9n^4 + 8n^3 + 15n$
 d. $2n^4 + 8n^3 + 15n + 3$

42. Multiply 1,987 × 0.05.
    a. 9.935
    b. 99.35
    c. 993.5
    d. 999.35

43. What is the product of the following expression?
$$(4x - 8)(5x^2 + x + 6)$$
    a. $20x^3 - 36x^2 + 16x - 48$

    b. $6x^3 - 41x^2 + 12x + 15$

    c. $204 + 11x^2 - 37x - 12$

    d. $2x^3 - 11x^2 - 32x + 20$

44. For the following similar triangles, what are the values of x and y (rounded to one decimal place)?

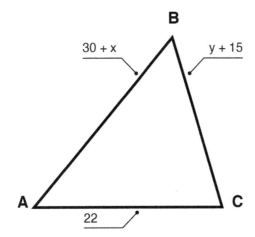
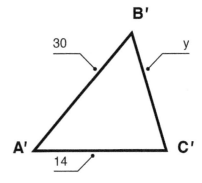

    a. $x = 16.5, y = 25.1$
    b. $x = 19.5, y = 24.1$
    c. $x = 17.1, y = 26.3$
    d. $x = 26.3, y = 17.1$

45. A solution needs 5 mL of saline for every 8 mL of medicine given. How much saline is needed for 45 mL of medicine?
    a. $\frac{225}{8}$ mL

    b. 72 mL

    c. 28 mL

    d. $\frac{45}{8}$ mL

46. If the volume of a sphere is 288π cubic meters, what are the radius and surface area of the same sphere?

    a. Radius 6 meters and surface area 144π square meters
    b. Radius 36 meters and surface area 144π square meters
    c. Radius 6 meters and surface area 12π square meters
    d. Radius 36 meters and surface area 12π square meters

47. The width of a rectangular house is 22 feet. What is the perimeter of this house if it has the same area as a house that is 33 feet wide and 50 feet long?

    a. 184 feet
    b. 200 feet
    c. 194 feet
    d. 206 feet

48. How much more area is covered by the rectangle than by the triangle?

8 inches

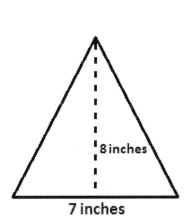

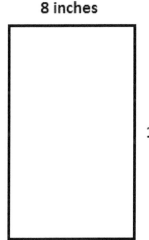

13 inches

8 inches

7 inches

    a. 48
    b. 21
    c. 104
    d. 76

49. How will the following number be written in standard form: $(1 \times 10^4) + (3 \times 10^3) + (7 \times 10^1) + (8 \times 10^0)$

    a. 137
    b. 13,078
    c. 1,378
    d. 8,731

50. If $3x = 6y = -2z = 24$, then what does $4xy + z$ equal?

    a. 116
    b. 130
    c. 84
    d. 108

51. Johnny earns $2334.50 from his job each month. He pays $1437 for monthly expenses. Johnny is planning a vacation in 3 months' time that he estimates will cost $1750 total. How much will Johnny have left over from three months' of saving once he pays for his vacation?

    a. $948.50

    b. $584.50

    c. $852.50

    d. $942.50

52. What is the value of the expression: $7^2 - 3 \times (4 + 2) + 15 \div 5$?

    a. 12.2

    b. 40.2

    c. 34

    d. 58.2

53. Gary is driving home to see his parents for Christmas. He travels at a constant speed of 60 miles per hour for a total of 350 miles. How many minutes will it take him to travel home if he takes a break for 10 minutes every 100 miles?

54. Kelly is selling cookies to raise money for the chorus. She has 500 cookies to sell. She sells 45% of the cookies to the sixth graders. At the second lunch, she sells 40% of what's left to the seventh graders. If she sells 60% of the remaining cookies to the eighth graders, how many cookies does Kelly have left at the end of all lunches?

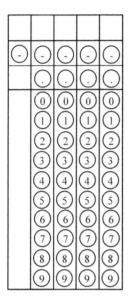

55. Find the value of x.

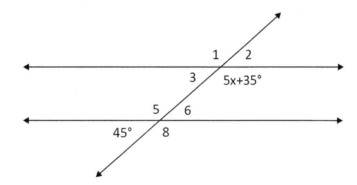

56. What is the value of the following expression?

$$\sqrt{8^2 + 6^2}$$

57. $864 \div 36 =$

Answer Explanations #1

Editing/Revising

1. D: The problem in the original sentence is that the second sentence is a dependent clause that cannot stand alone as a sentence; it must be attached to the main clause found in the first sentence. Because the main clause comes first, it does not need to be separated by a comma. However, if the dependent clause came first, then a comma would be necessary, which is why Choice *C* is incorrect. *A* and *B* also insert unnecessary commas into the sentence.

2. D: Choice *D* is correct because it uses correct parallel structure of plural nouns. *A* is incorrect because the word *shoe* should not be in singular form. Choice *B* is incorrect; semicolons are used in lists only if there is a list within a list. Choice *C* is incorrect because it pluralizes *makeup*, which is already in plural form.

3. C: The possessive form of the word "it" is "its." The contraction "it's" denotes "it is." Therefore, Choice *C* is correct. Choice *A* is incorrect because the end punctuation should be inside the quotes. Choice *B* is incorrect because removed the comma would create a run-on sentence. Distinguishes is the correct tense of the verb so Choice *D* is also incorrect.

4. B: Sentence 2 should be corrected to reflect the verb tense of the rest of the passage which is past tense. *Towards the end of his career, Carver returned to his first love of art.*

5. B: Move the sentence so that it comes before the preceding sentence. For this question, place the chosen sentence in each prospective choice's position. To move to the beginning of the paragraph is incorrect because the father "going crazy" doesn't logically begin the passage. Choice *C* is incorrect because the sentence in question is not a concluding sentence and does not transition smoothly into the second paragraph. Choice *D* is incorrect because the sentence doesn't necessarily need to be omitted since it logically follows the very first sentence in the passage.

6. D: Choice *D* is correct because "As it turns out" indicates a contrast from the previous sentiment, that the RV was a great purchase. Choice *A* is incorrect because "Unfortunately" indicates that the purchase was a negative one. Choice *B* is incorrect because the text indicates it *is* surprising that the RV was a great purchase because the author was skeptical beforehand. Choice *C* is incorrect because the transition "Furthermore" does not indicate a contrast.

7. B: This sentence calls for parallel structure. Choice *B* is correct because the verbs "wake," "eat," and "break" are consistent in tense and parts of speech. Choice *A* is incorrect because the word "eat" is present tense while the words "woke" and "broke" are in past tense. Choice *C* is incorrect because this turns the sentence into a question, which doesn't make sense within the context. Choice *D* is incorrect because it breaks tense with the rest of the passage. "Waking," "eating," and "breaking" are all present participles, and the context around the sentence is in past tense.

8. C: Choice *C* is correct because it is clear, concise, and fits within the context of the passage. Choice *A* is incorrect because the "hackers" in the sentence are meant to refer to the narrator and his family. Choice *B* is incorrect because it does not mention a solution being found and is therefore not specific enough. Choice *D* is incorrect because the meaning is eschewed by the helping verb "had to rejoice," and the sentence does not give enough detail as to what the problem entails.

9. C: This paragraph discusses positive aspects of traveling by RV including a sense of community with other RV owners. Sentence 11 raises the topic of the location of RV parks which does not directly relate to the rest of the paragraph.

10. B: Choice *B* is correct because there is no punctuation needed if a dependent clause ("while traveling across America") is located behind the independent clause ("it allowed us to share adventures"). Choice *A* is grammatically correct, but the wording is awkward and not the best choice for the combination. Choice *C* is incorrect because there are two dependent clauses connected and no independent clause, and a complete sentence requires at least one independent clause. Semicolons have the same function as periods: there must be an independent clause on either side of the semicolon. Choice *D* is incorrect because the dash simply interrupts the complete sentence.

11. C: The correct sentence to conclude this paragraph is Choice *C*, "Those are also memories that my siblings and I have now shared with our children." This choice relates to the memories the narrator made as he traveled by RV mentioned in sentence 15 and in the second paragraph. Choice *A* would make more sense as a conclusion to the first paragraph which discusses the father's aversion to nature. Choice *B* seems to belong in the second paragraph when the narrator discusses some of the times that things went wrong. The narrator never implies that he and his sibling never stay in hotels, but that they would have missed some of their adventures by doing so; therefore, Choice *D* is also incorrect.

Reading Comprehension

1. D: Choice *D* correctly summarizes Frost's theme of life's journey and the choices one makes. While Choice *A* can be seen as an interpretation, it is a literal one and is incorrect. Literal is not symbolic. Choice *B* presents the idea of good and evil as a theme, and the poem does not specify this struggle for the traveler. Choice *C* is a similarly incorrect answer. Love is not the theme.

2. C: Line 5 states, " Because it was grassy and wanted wear". The grass was taller and had not been trampled down by many travelers. The other choices are not lines that directly contribute to the idea that the second path was less traveled.

3. A: This phrase in the context of the poem seems to indicate that choosing a certain *way* leads to paths that might not have been found if a different *way* had been chosen. As each new path is taken, it gets harder to return to the original starting point.

4. A: There is only one traveler in the poem. We see this in the line "And be one traveler, long I stood." This indicates that the speaker is alone and traveling without a partner.

5. B: The time of day in the poem is morning. The line we see this in says, "And both that morning equally lay." This question relies on how carefully the passage was read.

6. B: The traveler took the second road. We see the traveler unsure at first about whether to take the first or second road. After they choose the second road, they contemplate saving "the first for another day," but they never take the first road in the poem.

41

7. D: It emphasizes Mr. Utterson's anguish in failing to identify Hyde's whereabouts. Context clues indicate that Choice *D* is correct because the passage provides great detail of Mr. Utterson's feelings about locating Hyde. Choice *A* does not fit because there is no mention of Mr. Lanyon's mental state. Choice *B* is incorrect; although the text does make mention of bells, Choice *B* is not the *best* answer overall. Choice *C* is incorrect because the passage clearly states that Mr. Utterson was determined, not unsure.

8. A: In the city. The word *city* appears in the passage several times, thus establishing the location for the reader.

9. B: It scares children. The passage states that the Juggernaut causes the children to scream. Choices *A* and *D* don't apply because the text doesn't mention either of these instances specifically. Choice *C* is incorrect because there is nothing in the text that mentions space travel.

10. B: To constantly visit. The mention of *morning*, *noon*, and *night* make it clear that the word *haunt* refers to frequent appearances at various locations. Choice *A* doesn't work because the text makes no mention of levitating. Choices *C* and *D* are not correct because the text makes mention of Mr. Utterson's anguish and disheartenment because of his failure to find Hyde but does not make mention of Mr. Utterson's feelings negatively affecting anyone else.

11. D: This is an example of alliteration. Choice *D* is the correct answer because of the repetition of the *L*-words. Hyperbole is an exaggeration, so Choice *A* doesn't work. No comparison is being made, so no simile or metaphor is being used, thus eliminating Choices *B* and *C*.

12. D: The speaker intends to continue to look for Hyde. Choices *A* and *B* are not possible answers because the text doesn't refer to any name changes or an identity crisis, despite Mr. Utterson's extreme obsession with finding Hyde. The text also makes no mention of a mistaken identity when referring to Hyde, so Choice *C* is also incorrect.

13. C: Gulliver becomes acquainted with the people and practices of his new surroundings. Choice *C* is the correct answer because it most extensively summarizes the entire passage. While Choices *A* and *B* are reasonable possibilities, they reference portions of Gulliver's experiences, not the whole. Choice *D* is incorrect because Gulliver doesn't express repentance or sorrow in this particular passage.

14. A: Principal refers to *chief* or *primary* within the context of this text. Choice *A* is the answer that most closely aligns with this answer. Choices *B* and *D* make reference to a helper or followers while Choice *C* doesn't meet the description of Gulliver from the passage.

15. C: One can reasonably infer that Gulliver is considerably larger than the children who were playing around him because multiple children could fit into his hand. Choice *B* is incorrect because there is no indication of stress in Gulliver's tone. Choices *A* and *D* aren't the best answer because though Gulliver seems fond of his new acquaintances, he didn't travel there with the intentions of meeting new people or to express a definite love for them in this particular portion of the text.

16. C: The emperor made a *definitive decision* to expose Gulliver to their native customs. In this instance, the word *mind* was not related to a vote, question, or cognitive ability.

17. A: Choice *A* is correct. This assertion does *not* support the fact that games are a commonplace event in this culture because it mentions conduct, not games. Choices *B*, *C*, and *D* are incorrect because these do support the fact that games were a commonplace event.

18. B: Choice *B* is the only option that mentions the correlation between physical ability and leadership positions. Choices *A* and *D* are unrelated to physical strength and leadership abilities. Choice *C* does not make a deduction that would lead to the correct answer—it only comments upon the abilities of common townspeople.

19. B: It denotes a period of time. It is apparent that Lincoln is referring to a period of time within the context of the passage because of how the sentence is structured with the word *ago*.

20. C: Lincoln's reference to *the brave men, living and dead, who struggled here,* proves that he is referring to a battlefield. Choices *A* and *B* are incorrect, as a *civil war* is mentioned and not a war with France or a war in the Sahara Desert. Choice *D* is incorrect because it does not make sense to consecrate a President's ground instead of a battlefield ground for soldiers who died during the American Civil War.

21. D: Abraham Lincoln is the former president of the United States, and he references a "civil war" during his address.

22. A: The audience should consider the death of the people that fought in the war as an example and perpetuate the ideals of freedom that the soldiers died fighting for. Lincoln doesn't address any of the topics outlined in Choices *B, C,* or *D*. Therefore, Choice *A* is the correct answer.

23. D: Choice *D* is the correct answer because of the repetition of the word *people* at the end of the passage. Choice *A, antimetabole,* is the repetition of words in a succession. Choice *B, antiphrasis,* is a form of denial of an assertion in a text. Choice *C, anaphora,* is the repetition that occurs at the beginning of sentences.

24. A: Choice *A* is correct because Lincoln's intention was to memorialize the soldiers who had fallen as a result of war as well as celebrate those who had put their lives in danger for the sake of their country. Choices *B* and *D* are incorrect because Lincoln's speech was supposed to foster a sense of pride among the members of the audience while connecting them to the soldiers' experiences.

25. C: In lines 6 and 7, it is stated that avarice can prevent a man from being necessitously poor, but too timorous, or fearful, to achieve real wealth. According to the passage, avarice does tend to make a person very wealthy. The passage states that oppression, not avarice, is the consequence of wealth. The passage does not state that avarice drives a person's desire to be wealthy.

26. D: Paine believes that the distinction that is beyond a natural or religious reason is between king and subjects. He states that the distinction between good and bad is made in heaven. The distinction between male and female is natural. He does not mention anything about the distinction between humans and animals.

27. A: The passage states that the Heathens were the first to introduce government by kings into the world. The quiet lives of patriarchs came before the Heathens introduced this type of government. It was Christians, not Heathens, who paid divine honors to living kings. Heathens honored deceased kings. Equal rights of nature are mentioned in the paragraph, but not in relation to the Heathens.

28. B: Paine asserts that a monarchy is against the equal rights of nature and cites several parts of scripture that also denounce it. He doesn't say it is against the laws of nature. Because he uses scripture to further his argument, it is not despite scripture that he denounces the monarchy. Paine addresses the law by saying the courts also do not support a monarchical government.

29. A: To be *idolatrous* is to worship idols or heroes, in this case, kings. It is not defined as being deceitful. While idolatry is considered a sin, it is an example of a sin, not a synonym for it. Idolatry may have been considered illegal in some cultures, but it is not a definition for the term.

30. A: The tone is exasperated. While contemplative is an option because of the inquisitive nature of the text, Choice *A* is correct because the speaker is annoyed by the thought of being included when he felt that the fellow members of his race were being excluded. The speaker is not nonchalant, nor accepting of the circumstances which he describes.

31. C: Choice *C*, *contented*, is the only word that has different meaning. Furthermore, the speaker expresses objection and disdain throughout the entire text.

32. B: The main focus is to address the feelings of exclusion expressed by African Americans after the establishment of the Fourth of July holiday. While the speaker makes biblical references, it is not the main focus of the passage, thus eliminating Choice *A* as an answer. The passage also makes no mention of wealthy landowners and doesn't speak of any positive response to the historical events, so Choices *C* and *D* are not correct.

33: D: Choice *D* is the correct answer because it clearly makes reference to justice being denied.

34: D: It is an example of hyperbole. Choices *A* and *B* are unrelated. Assonance is the repetition of sounds and commonly occurs in poetry. Parallelism refers to two statements that correlate in some manner. Choice *C* is incorrect because amplification normally refers to clarification of meaning by broadening the sentence structure, while hyperbole refers to a phrase or statement that is being exaggerated.

35: C: Display the equivocation of the speaker and those that he represents. Choice *C* is correct because the speaker is clear about his intention and stance throughout the text. Choice *A* could be true, but the words "common text" is arguable. Choice *B* is also partially true, as another group of people affected by slavery are being referenced. However, the speaker is not trying to convince the audience that injustices have been committed, as it is already understood there have been injustices committed. Choice *D* is also close to the correct answer, but it is not the *best* answer choice possible.

36. D: The use of "I" could have all of the effects for the reader; it could serve to have a "hedging" effect, allow the reader to connect with the author in a more personal way, and cause the reader to empathize more with the egrets. However, it doesn't distance the reader from the text, thus eliminating Choice *D*.

37. C: The quote provides an example of a warden protecting one of the colonies. Choice *A* is incorrect because the speaker of the quote is a warden, not a hunter. Choice *B* is incorrect because the quote does not lighten the mood but shows the danger of the situation between the wardens and the hunters. Choice *D* is incorrect because there is no humor found in the quote.

38. D: A *rookery* is a colony of breeding birds. Although *rookery* could mean Choice *A*, houses in a slum area, it does not make sense in this context. Choices *B* and *C* are both incorrect, as this is not a place for hunters to trade tools or for wardens to trade stories.

39. B: An important bird colony. The previous sentence is describing "twenty colonies" of birds, so what follows should be a bird colony. Choice *A* may be true, but we have no evidence of this in the text.

Choice *C* does touch on the tension between the hunters and wardens, but there is no official "Bird Island Battle" mentioned in the text. Choice *D* does not exist in the text.

40. D: To demonstrate the success of the protective work of the Audubon Association. The text mentions several different times how and why the association has been successful and gives examples to back this fact. Choice *A* is incorrect because although the article, in some instances, calls certain people to act, it is not the purpose of the entire passage. There is no way to tell if Choices *B* and *C* are correct, as they are not mentioned in the text.

41. C: To have a better opportunity to hunt the birds. Choice *A* might be true in a general sense, but it is not relevant to the context of the text. Choice *B* is incorrect because the hunters are not studying lines of flight to help wardens, but to hunt birds. Choice *D* is incorrect because nothing in the text mentions that hunters are trying to build homes underneath lines of flight of birds for good luck.

42. A: It introduces certain insects that transition from water to air. Choice *B* is incorrect because although the passage talks about gills, it is not the central idea of the passage. Choices *C* and *D* are incorrect because the passage does not "define" or "invite," but only serves as an introduction to stoneflies, dragonflies, and mayflies and their transition from water to air.

43. C: The act of shedding part or all of the outer shell. Choices *A, B,* and *D* are incorrect.

44. B: The first paragraph serves as a contrast to the second. Notice how the first paragraph goes into detail describing how insects are able to breathe air. The second paragraph acts as a contrast to the first by stating "[i]t is of great interest to find that, nevertheless, a number of insects spend much of their time under water." Watch for transition words such as "nevertheless" to help find what type of passage you're dealing with.

45. C: The stage of preparation in between molting is acted out in the water, while the last stage is in the air. Choices *A, B,* and *D* are all incorrect. *Instars* is the phase between two periods of molting, and the text explains when these transitions occur.

46. C: The author's tone is informative and exhibits interest in the subject of the study. Overall, the author presents us with information on the subject. One moment where personal interest is depicted is when the author states, "It is of great interest to find that, nevertheless, a number of insects spend much of their time under water."

Math

1. B: First, subtract 4 from each side. This yields $6t = 12$. Now, divide both sides by 6 to obtain $t = 2$.

2. B: To be directly proportional means that $y = mx$. If x is changed from 5 to 20, the value of x is multiplied by 4. Applying the same rule to the y-value, also multiply the value of y by 4. Therefore:

$$y = 12$$

3. B: From the slope-intercept form, $y = mx + b$, it is known that b is the y-intercept, which is 1. Compute the slope as $\frac{2-1}{1-0} = 1$, so the equation should be $y = x + 1$.

4. A: Each bag contributes $4x + 1$ treats. The total treats will be in the form $4nx + n$ where n is the total number of bags. The total is in the form $60x + 15$, from which it is known $n = 15$.

5. B: $350,000: Since the total income is $500,000, then a percentage of that can be found by multiplying the percent of Audit Services as a decimal, or 0.40, by the total of 500,000. This answer is found from the equation:

$$500000 \times 0.4 = 200000$$

The total income from Audit Services is $200,000.

For the income received from Taxation Services, the following equation can be used:

$$500000 \times 0.3 = 150000$$

The total income from Audit Services and Taxation Services is $150,000 + 200,000 = 350,000$.

Another way of approaching the problem is to calculate the easy percentage of 10%, then multiply it by 7, because the total percentage for Audit and Taxation Services was 70%. 10% of 500,000 is 50,000. Then multiplying this number by 7 yields the same income of $350,000.

6. A: Finding the roots means finding the values of x when y is zero. The quadratic formula could be used, but in this case, it is possible to factor by hand, since the numbers -1 and 2 add to 1 and multiply to -2. So, factor $x^2 + x - 2 = (x - 1)(x + 2) = 0$, then set each factor equal to zero. Solving for each value gives the values $x = 1$ and $x = -2$.

7. C: To find the y-intercept, substitute zero for x, which gives us:

$$y = 0^{\frac{5}{3}} + (0 - 3)(0 + 1)$$

$$0 + (-3)(1) = -3$$

8. D: The slope from this equation is 50, and it is interpreted as the cost per gigabyte used. Since the g-value represents number of gigabytes and the equation is set equal to the cost in dollars, the slope relates these two values. For every gigabyte used on the phone, the bill goes up 50 dollars.

9. A: Simplify this to:

$$(4x^2y^4)^{\frac{3}{2}} = 4^{\frac{3}{2}}(x^2)^{\frac{3}{2}}(y^4)^{\frac{3}{2}}$$

$$4^{\frac{3}{2}} = (\sqrt{4})^3 = 2^3 = 8$$

For the other, recall that the exponents must be multiplied; this yields:

$$8x^{2 \cdot \frac{3}{2}}y^{4 \cdot \frac{3}{2}} = 8x^3y^6$$

10. B: Start by squaring both sides to get $1 + x = 16$. Then subtract 1 from both sides to get $x = 15$.

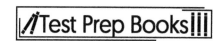

11. C: Multiply both sides by x to get $x + 2 = 2x$, which simplifies to $-x = -2$, or $x = 2$.

12. A: Parallel lines have the same slope. The slope of C can be seen to be 1/3 by dividing both sides by 3. The others are in standard form $Ax + By = C$, for which the slope is given by $\frac{-A}{B}$. The slope of A is 3, the slope of B is 4. The slope of D is 1.

13. D: Denote the width as w and the length as l. Then, $l = 3w + 5$. The perimeter is $2w + 2l = 90$. Substituting the first expression for l into the second equation yields $2(3w + 5) + 2w = 90$, or $8w = 80$, so $w = 10$. Putting this into the first equation, it yields:

$$l = 3(10) + 5 = 35$$

14. A: Lining up the given scores provides the following list: 60, 75, 80, 85, and one unknown. Because the median needs to be 80, it means 80 must be the middle data point out of these five. Therefore, the unknown data point must be the fourth or fifth data point, meaning it must be greater than or equal to 80. The only answer that fails to meet this condition is 60.

15. B: If 60% of 50 workers are women, then there are 30 women working in the office. If half of them are wearing skirts, then that means 15 women wear skirts. Since none of the men wear skirts, this means there are 15 people wearing skirts.

16. A: Let the unknown score be x. The average will be:

$$\frac{(5 \times 50) + (4 \times 70) + x}{10} = \frac{530 + x}{10} = 55$$

Multiply both sides by 10 to get $530 + x = 550$, or $x = 20$.

17. D: For manufacturing costs, there is a linear relationship between the cost to the company and the number produced, with a y-intercept given by the base cost of acquiring the means of production, and a slope given by the cost to produce one unit. In this case, that base cost is $50,000, while the cost per unit is $40. So:

$$y = 40x + 50,000$$

18. C: A die has an equal chance for each outcome. Since it has six sides, each outcome has a probability of $\frac{1}{6}$. The chance of a 1 or a 2 is therefore:

$$\frac{1}{6} + \frac{1}{6} = \frac{1}{3}$$

19. A: The slope is given by:

$$m = \frac{y_2 - y_1}{x_2 - x_1} = \frac{0 - 4}{0 - (-3)} = -\frac{4}{3}$$

20. B: An equilateral triangle has three sides of equal length, so if the total perimeter is 18 feet, each side must be 6 feet long. A square with sides of 6 feet will have an area of $6^2 = 36$ square feet.

21. A: 3.6. Divide 3 by 5 to get 0.6 and add that to the whole number 3, to get 3.6. An alternative is to incorporate the whole number 3 earlier on by creating an improper fraction:

$$\frac{18}{5}$$

Then dividing 18 by 5 to get 3.6.

22. C: 9 Cars. The average is calculated by adding up each month's sales and dividing the sum by the total number of months in the time period. Dealer 1 sold 2 cars in July, 12 in August, 8 in September, 6 in October, 10 in November, and 15 in December. The sum of these sales is:

$$2 + 12 + 8 + 6 + 10 + 15 = 53 \text{ cars}$$

To find the average, this sum is divided by the total number of months, which is 6. When 53 is divided by 6, it yields 8.8333... Since cars are sold in whole numbers, the answer is rounded to 9 cars.

23. A: 6 boxes. The team needs a total of $270, and each box earns them $3. Therefore, the total number of boxes needed to be sold is $270 \div 3$, which is 90. With 15 people on the team, the total of 90 can be divided by 15, which equals 6. This means that each member of the team needs to sell 6 boxes for the team to raise enough money to buy new uniforms.

24. C: 30. A complete circle measures 360 degrees. This circle is broken up into 4 different parts with different measures for each part. Adding these parts should give a total of 360 degrees. The equation generated from this diagram is:

$$4x + 5x + x + 2x = 360$$

Collecting like terms gives the equation $12x = 360$, which can be solved by dividing by 12 to give $x = 30$. The value of x in the diagram is 30.

25. D: $\frac{1}{12}$. The probability of picking the winner of the race is:

$$\frac{1}{4}\left(\frac{number\ of\ favorable\ outcomes}{number\ of\ total\ outcomes}\right)$$

Assuming the winner was picked on the first selection, three horses remain from which to choose the runner-up (these are dependent events). Therefore, the probability of picking the runner-up is $\frac{1}{3}$. To determine the probability of multiple events, the probability of each event is multiplied:

$$\frac{1}{4} \times \frac{1}{3} = \frac{1}{12}$$

26. C: A full rotation is 360 degrees. Taking the total degrees that the figure skater spins and dividing by 360 yields 6.25. She spins 6 total times and then one quarter of a turn more. This quarter of a turn to her right means she ends up facing East.

27. C: The sum of all angles in a polygon with n sides is found by the expression $(n - 2) \times 180$. Since this polygon has 5 sides, then the total degrees of the interior angles can be found using the equation:

$$(5 - 2) \times 180 = 540$$

Adding up each of the given angles yields a total of:

$$111 + 113 + 92 + 128 = 444 \text{ degrees}$$

Taking the total of 540 degrees and subtracting the given sum of 444 degrees gives a missing value of 96 degrees for the measure of angle P.

28. A: The expression in the denominator can be factored into the two binomials:

$$(x - 4)(x - 2)$$

Once the expression is rewritten as $\frac{x-4}{(x-4)(x-2)}$, the values of $x = 4$ and $x = 2$ result in a denominator with a value of 0. Since 0 cannot be in the denominator of a fraction, the expression is undefined at the values of $x = 2, 4$.

29. A: The total number of coins in the jar is 86, which is the sum of all the coins. The probability of Nina choosing a coin other than a penny or a dime can be found by calculating the total of quarters and nickels. This total is 31. Taking 31 and dividing it by 86 gives the probability of choosing a coin that is not a penny or a dime. This decimal found from the fraction $\frac{31}{86}$ is 0.36.

30. C: For the first car, the trip will be 450 miles at 18 miles to the gallon. The total gallons needed for this car will be:

$$450 \div 18 = 25$$

For the second car, the trip will be 450 miles at 25 miles to the gallon, or $450 \div 25 = 18$, which will require 18 gallons of gas. Adding these two amounts of gas gives a total of 43 gallons of gas. If the gas costs $2.49 per gallon, the cost of the trip for both cars is:

$$43 \times \$2.49 = \$107.07$$

31. C: $\frac{4}{3}$

Set up the problem and find a common denominator for both fractions.

$$\frac{14}{33} + \frac{10}{11}$$

Multiply each fraction across by 1 to convert to a common denominator

$$\frac{14}{33} \times \frac{1}{1} + \frac{10}{11} \times \frac{3}{3}$$

Once over the same denominator, add across the top. The total is over the common denominator.

$$\frac{14 + 30}{33} = \frac{44}{33}$$

Reduce by dividing both numerator and denominator by 11.

$$\frac{44 \div 11}{33 \div 11} = \frac{4}{3}$$

32. B: 128

This question involves the percent formula.

$$\frac{32}{x} = \frac{25}{100}$$

We multiply the diagonal numbers, 32 and 100, to get 3,200. Dividing by the remaining number, 25, gives us 128.

The percent formula does not have to be used for a question like this. Since 25% is ¼ of 100, you know that 32 needs to be multiplied by 4, which yields 128.

33. D: The first step in solving this equation is to collect like terms on the left side of the equation. This yields the new equation:

$$-4 + 8x = 8 - 10x$$

The next step is to move the x-terms to one side by adding 10 to both sides, making the equation:

$$-4 + 18x = 8$$

Then the -4 can be moved to the right side of the equation to form:

$$18x = 12$$

Dividing both sides of the equation by 18 gives a value of 0.67, or $\frac{2}{3}$.

34. B: This triangle can be labeled as a right triangle because it has a right-angle measure in the corner. The Pythagorean Theorem can be used here to find the missing side lengths. The Pythagorean Theorem states that $a^2 + b^2 = c^2$, where a and b are side lengths and c is the hypotenuse. The hypotenuse, c, is equal to 35, and 1 side, a, is equal to 21. Plugging these values into the equation forms:

$$21^2 + b^2 = 35^2$$

Squaring both given numbers and subtracting them yields the equation:

$$b^2 = 784$$

Taking the square root of 784 gives a value of 28 for b. In the equation, b is the same as the missing side length x.

35. A: The first step is to find the equation of the line that is perpendicular to $y = 2x - 3$ and passes through the point $(0, 5)$. The slope of a perpendicular line is found by the negative reciprocal of 2, which is $-\frac{1}{2}$. The y-intercept is the value of y when $x = 0$, so the y-intercept is 5. The new equation is:

$$y = -\frac{1}{2}x + 5$$

In order to find which points lie on the new line, the values of x and y can be substituted into the equation to determine if they form a true statement. For A, the equation $4 = -\frac{1}{2}(2) + 5$ makes a true statement, so the point $(2, 4)$ lies on the lines.

For B, the equation $7 = -\frac{1}{2}(-2) + 5$ makes the statement $7 = 6$, which is not a true statement. Therefore, B is not a point that lies on the line. For C, the equation $-3 = -\frac{1}{2}(4) + 5$ becomes $-3 = 3$ which is not a true statement, so the point $(4, -3)$ is not on the line.

For the last point in D, the equation $10 = -\frac{1}{2}(-6) + 5$ makes the statement $10 = 8$. This is not a true statement, so the point $(-6, 10)$ does not lie on the line.

36. B: To factor $x^2 + 4x + 4$, the numbers needed are those that add to 4 and multiply to 4. Therefore, both numbers must be 2, and the expression factors to:

$$x^2 + 4x + 4 = (x + 2)^2$$

Similarly, the expression factors to $x^2 - x - 6 = (x - 3)(x + 2)$, so that they have $x + 2$ in common.

37. D: This problem involves a composition function, where one function is plugged into the other function. In this case, the $f(x)$ function is plugged into the $g(x)$ function for each x-value. The composition equation becomes:

$$g(f(x)) = 2^3 - 3(2^2) - 2(2) + 6$$

Simplifying the equation gives the answer:

$$g(f(x)) = 8 - 3(4) - 2(2) + 6$$

$$8 - 12 - 4 + 6 = -2$$

38. B: 100 cm is equal to 1 m. 1.3 divided by 100 is 0.013. Therefore, 1.3 cm is equal to 0.013 mm. Because 1 cm is equal to 10 mm, 1.3 cm is equal to 13 mm.

39. B: 725

Set up the division problem.

$$1.4 \overline{)1015}$$

Move the decimal over one place to the right in both numbers.

$$14 \overline{)10150}$$

14 does not go into 1 or 10 but does go into 101 so start there.

```
        7 2 5
14)1 0 1 5 0
  - 9 8
    3 5
  - 2 8
      7 0
    - 7 0
        0
```

The result is 725.

40. D: This system of equations involves one quadratic function and one linear function, as seen from the degree of each equation. One way to solve this is through substitution. Solving for y in the second equation yields $y = x + 2$. Plugging this equation in for the y of the quadratic equation yields:

$$x^2 - 2x + x + 2 = 8$$

Simplifying the equation, it becomes:

$$x^2 - x + 2 = 8$$

Setting this equal to zero and factoring, it becomes:

$$x^2 - x - 6 = 0 = (x - 3)(x + 2)$$

Solving these two factors for x gives the zeros $x = 3, -2$. To find the y-value for the point, each number can be plugged in to either original equation. Solving each one for y yields the points $(3, 5)$ and $(-2, 0)$.

41. D: The expression is simplified by collecting like terms. Terms with the same variable and exponent are like terms, and their coefficients can be added.

42. B: 99.35. Set up the problem, with the larger number on top. Multiply as if there are no decimal places. Add the answer rows together. Count the number of decimal places that were in the original numbers (2).

Place the decimal in that many spots from the right for the final solution.

43. A: Finding the product means distributing one polynomial to the other so that each term in the first is multiplied by each term in the second. Then, like terms can be collected. Multiplying the factors yields the expression:

$$20x^3 + 4x^2 + 24x - 40x^2 - 8x - 48$$

Collecting like terms means adding the x^2 terms and adding the x terms. The final answer after simplifying the expression is:

$$20x^3 - 36x^2 + 16x - 48$$

44. C: Because the triangles are similar, the lengths of the corresponding sides are proportional. Therefore:

$$\frac{30 + x}{30} = \frac{22}{14} = \frac{y + 15}{y}$$

This results in the equation $14(30 + x) = 22 \times 30$ which, when solved, gives $x = 17.1$. The proportion also results in the equation $14(y + 15) = 22y$ which, when solved, gives $y = 26.3$.

45. A: Every 8 ml of medicine requires 5 mL. The 45 mL first needs to be split into portions of 8 mL. This results in $\frac{45}{8}$ portions. Each portion requires 5 mL. Therefore, the following is necessary:

$$\frac{45}{8} \times 5 = \frac{45 \times 5}{8} = \frac{225}{8} \text{ mL}$$

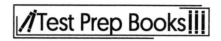

46. A: Because the volume of the given sphere is 288π cubic meters, this means:

$$\frac{4}{3}\pi r^3 = 288\pi$$

This equation is solved for r to obtain a radius of 6 meters. The formula for the surface area of a sphere is $4\pi r^2$, so if $r = 6$ in this formula, the surface area is 144π square meters.

47. C: First, find the area of the second house. The area is:

$$A = l \, x \, w = 33 \times 50 = 1,650 \text{ square feet}$$

Then use the area formula to determine what length gives the first house an area of 1,650 square feet. So:

$$1,650 = 22 \times l$$

$$l = \frac{1,650}{22} = 75 \text{ feet}$$

Then, use the formula for perimeter to get:

$$75 + 75 + 22 + 22 = 194 \text{ feet}$$

48. D: First, find the area of both figures. The area of the triangle is:

$$\frac{1}{2}(7) \times 8 = 28 \text{ square inches}$$

The area of the rectangle is:

$$13 \times 8 = 104 \text{ square inches}$$

To find how much more area is covered by the square, the following equation can be used:

$$104 - 28 = 76$$

49. B: 13,078. The power of 10 by which a digit is multiplied corresponds with the number of zeros following the digit when expressing its value in standard form. Therefore:

$$(1 \times 10^4) + (3 \times 10^3) + (7 \times 10^1) + (8 \times 10^0)$$

$$10,000 + 3,000 + 70 + 8 = 13,078$$

50. A: First solve for *x, y,* and *z.* So:

$$3x = 24$$

$$x = 8$$

$$6y = 24$$

$$y = 4$$

$$-2z = 24$$

$$z = -12$$

This means the equation would be $4(8)(4) + (-12)$, which equals 116.

51. D: First, subtract $1437 from $2334.50 to find Johnny's monthly savings; this equals $897.50. Then, multiply this amount by 3 to find out how much he will have (in three months) before he pays for his vacation: this equals $2692.50. Finally, subtract the cost of the vacation ($1750) from this total to find how much Johnny will have left: $942.50.

52. C: 34. When performing calculations consisting of more than one operation, the order of operations should be followed: *Parenthesis, Exponents, Multiplication/Division, Addition/Subtraction*.

Parenthesis:

$$7^2 - 3 \times (4 + 2) + 15 \div 5$$

$$7^2 - 3 \times (6) + 15 \div 5$$

Exponents:

$$7^2 - 3 \times 6 + 15 \div 5$$

$$49 - 3 \times 6 + 15 \div 5$$

Multiplication/Division (from left to right):

$$49 - 3 \times 6 + 15 \div 5$$

$$49 - 18 + 3$$

Addition/Subtraction (from left to right):

$$49 - 18 + 3 = 34$$

53.

		3	8	0

380 miles. To find the total driving time, the total distance of 350 miles can be divided by the constant speed of 60 miles per hour. This yields a time of 5.8333 hours, which is then rounded. Once the driving time is computed, the break times need to be found. If Gary takes a break for 10 minutes every hour, he will take 3 breaks on his trip. This will yield a total of 30 minutes of break time. Since the answer is needed in minutes, 5.8333 can be converted to minutes by multiplying by 60, giving a driving time of 350 minutes. Adding the break time of 30 minutes to the driving time of 350 minutes gives a total travel time of 380 minutes.

54.

66 Cookies. If the sixth graders bought 45% of the cookies, the number they bought is found by multiplying 0.45 by 500. They bought 225 cookies. The number of cookies left is:

$$500 - 225 = 275$$

During the second lunch, the seventh graders bought 40% of the cookies, which is found by multiplying 0.40 by the remaining 275 cookies. The seventh graders bought 110 cookies. This leaves 165 cookies to sell to the eighth graders. If they bought 60% of the remaining cookies, then they bought 99 cookies. Subtracting 99 from 165 cookies leaves Kelly with 66 cookies remaining after the three lunches.

55.

Answer grid: **2 0**

20. Because these are 2 parallel lines cut by a transversal, the angle with a measure of 45 degrees is equal to the measure of angle 6. Angle 6 and the angle labeled $5x + 35$ are supplementary to one another. The sum of these angles should be 180, so the following equation can be generated:

$$5x + 35 + 45 = 180$$

Solving for x, the sum of 35 and 45 is 80, which is then subtracted from 180 to yield 100. Dividing 100 by 5 gives the value of x, which is 20.

56.

Answer grid: **1 0**

10. 8 squared is 64, and 6 squared is 36. These should be added together to get:

$$64 + 36 = 100$$

Then, the last step is to find the square root of 100 which is 10.

57.

24; The long division would be completed as follows:

```
        24
   36|864
    - 72↓
      144
```

SHSAT Practice Test #2

Editing/Revising

Editing/Revising Part A

1. Read this paragraph.

> (1) *Romeo and Juliet*'s a well known tragedy written by William Shakespeare. (2) The drama depicts a story of two teenagers from families of different social classes. (3) Romeo and Juliet fall in love at a masquerade ball and, soon after, run away to get married. (4) Tragically, their love story quickly ends when a series of events leads to their deaths.

How should the paragraph be revised?
   a. Sentence 1: Change *Juliet's* to **Juliet** *is* AND *well known* to **well-known**.
   b. Sentence 2: Change *two* to **too** AND *from* to **of**.
   c. Sentence 3: Change *fall* to **fell** AND *run* to **ran**.
   d. Sentence 4: Delete the comma after *Tragically* AND change *ends* to **end**.

2. Read this paragraph.

> (1) The respiratory system is vital to the human body. (2) When air is inhaled the lungs extract oxygen from the air and send it into the blood. (3) The heart then pumps the blood through a series of veins and arteries to deliver the oxygen to the body. (4) Once the oxygen has been removed from the blood, carbon dioxide is exhaled.

Which sentence should be revised to correct an error in sentence structure?
   a. Sentence 1
   b. Sentence 2
   c. Sentence 3
   d. Sentence 4

3. Read these sentences.

> People who text while driving look away from the road for a minimum of five seconds.

> Driving fifty-five miles per hour, it takes at least five seconds to travel the length of a football field.

What is the best way to combine the sentences to clarify the relationship between the ideas?
   a. It takes five seconds to travel the length of a football field, which is the same amount of time it takes to send a text while driving.
   b. It is as dangerous to text and drive as it is to drive fifty-five miles per hour, the length of a football field.
   c. Sending a text while driving is equivalent to taking your eyes off the road while driving the length of a football field at fifty-five miles per hour.
   d. It takes five seconds to travel the length of a football field without looking, so texting while driving only takes about five seconds.

4. Read this sentence.

> Mr. Wilkinson arrived late to work due to the fact that he felt it necessary to stop and purchase an assortment of breakfast items for the class as a reward for their excellent test scores last week.

Which revision uses the most concise language?

a. Mr. Wilkinson arrived late to work due to the fact that he felt it necessary to stop and get breakfast for the class.
b. He felt it necessary to stop and purchase breakfast items for the class as a reward for their excellent test scores last week.
c. Mr. Wilkinson stopped to purchase breakfast for the class as a reward.
d. Mr. Wilkinson was late to work because he stopped to buy breakfast for his class as a reward for their good test scores last week.

5. Read this paragraph.

> (1) The American Flag is a symbol of unity strength and courage for the United States. (2) The flag design consists of thirteen red and white horizontal stripes and fifty white stars against a blue background. (3) The original design of the flag only had thirteen stars to represent the original thirteen colonies. (4) However, after the Civil War, the design was revised to display fifty white stars to represent the fifty states.

How should the paragraph be revised?

a. Sentence 1: Change *Flag* to **flag** AND *unity strength and courage* to **unity, strength, and courage**.
b. Sentence 2: Insert a comma after *red* AND insert a comma after *stripes*.
c. Sentence 3: Change *had* to **has** AND *represent* to **represented**.
d. Sentence 4: Remove the comma after *War* AND insert a comma after *stars*.

6. Read these sentences.

> A penguin's bones are much denser than those of other birds.

> Penguins are birds, but they float and dive instead of flying.

What is the best way to combine the sentences to clarify the relationship between the ideas?

a. Although penguins are birds, they can't fly because their bones are too heavy.
b. Because penguins have heavy bones, they cannot fly, but they can float and dive.
c. Penguins are heavier than other birds, which helps them to float and dive.
d. Penguins have adapted to swimming instead of flying because they live close to the water.

7. Read this paragraph.

> (1) The Statue of Liberty is one of the most well-known statues in the world, standing 305 feet, 6 inches tall on Liberty Island. (2) The statue is sheeted in gold and patina and weighs 450,000 pounds. (3) The seven rays on Lady Libertys crown, representing the seven continents, weighs as much as 150 pounds and is up to 9 feet long. (4) The tablet she holds in her left hand is inscribed with America's date of independence, July 4, 1776.

How should the paragraph be revised?

a. Sentence 1: Change *Island* to **island** AND *well-known* to **iconic**.
b. Sentence 2: Change *gold and patina and weighs* to **gold, patina, and weighs** AND *pounds* to **lbs**.
c. Sentence 3: Change *Libertys* to **Liberty's** AND *weighs* to **weigh**.
d. Sentence 4: Change *inscribed* to **enscribed** AND *independence* to **Independence**.

Editing/Revising Part B

Read the text below and answer the questions following it.

(1) The country of Japan is a chain of islands located along the coast of Asia. (2) Japan is made up of four major islands, with Honshu being the largest. (3) Honshu is also the seventh largest island in the world. (4) Underneath Japan, there are many volcanoes, some active and some dormant, but only about 108 of them are active today.

(5) There are many exciting places to visit in Japan. (6) Osaka is home to Dontobori, the main entertainment district. (7) The popular sport of sumo wrestling is often hosted in the town's arena. (8) People from all over Japan ride the Shinkansen or bullet train to Osaka to watch the events. (9) Hakone is famous for black eggs. (10) The town of Hakone sits on top of a mountain near Mount Fuji, the tallest mountain in Japan. (11) The eggs are boiled in a hot spring on the mountainside. (12) The natural water turns the shell of the egg black; however, it does not change the taste of it. (13) Tokyo is the capital city of Japan. (14) The town of Kyoto is full of shrines and temples, some lined with gold. (15) It is also a well-known place for geisha sightings.

(16) Traditional Japanese food consists mostly of fish and rice. (17) Sushi is raw fish and vegetables rolled in rice and wrapped in seaweed. (18) It is typically dipped in a sauce made of soy sauce and wasabi paste. (19) Tempura is a light, crispy batter used to fry fish, chicken, or vegetables. (20) Soba is another traditional Japanese dish. (21) It is a type of noodle that is oftentimes eaten cold and served in a broth made with soy sauce.

8. Which transition sentence should be added to the beginning of sentence 16?
   a. Typical Japanese lunch boxes are called *bentos* and usually consist of foods such as dried fruit, rice, and sushi.
   b. Japanese cuisine is much different from American cuisine.
   c. Yakisoba is a traditional Japanese noodle dish served with slices of beef.
   d. Gyoza is a traditional Japanese dish that many Americans refer to as *pot stickers*.

9. Which sentence should be moved to follow sentence 8 to correct the flow of information in the paragraph?
   a. Sentence 5
   b. Sentence 7
   c. Sentence 10
   d. Sentence 15

10. Which revision of sentence 4 uses the most precise language?
    a. Japan sits just above 108 active volcanos.
    b. Japan has many volcanos, with only 108 being active.
    c. Active volcanos are common in Japan, especially the biggest 108.
    d. Japan lies on a chain of volcanos, and more than 108 of them are active.

11. Which sentence should be revised to correct an error in sentence structure?
    a. Sentence 4
    b. Sentence 8
    c. Sentence 12
    d. Sentence 14

Reading Comprehension

Questions 1–6 are based on the following poem, "To Waken an Old Lady" by William Carlos Williams:

Old age is
a flight of small
cheeping birds
skimming
bare trees   5
above a snow glaze.
Gaining and failing
they are buffeted
by a dark wind—
But what?   10
On harsh weedstalks
the flock has rested,
the snow
is covered with broken
seedhusks   15
and the wind tempered
by a shrill
piping of plenty.

1. This poem uses which of the following literary techniques typical of its time period?
    a. Meter
    b. Anaphora
    c. Imagery
    d. Synecdoche

2. This poem comes out of which of the following literary periods?
    a. Romanticism
    b. Modernism
    c. Postmodernism
    d. Confessional poetry

3. Which of the following provides the best analysis of the poem?
   a. The poem acts as an extended metaphor of old age. Its juxtaposed imagery suggests the frailty of life ("Gaining and failing / they are buffeted / by a dark wind") alongside the fullness of life and its "piping of plenty."
   b. The poem describes a flock of birds and their relationship with nature. They rest "On harsh weedstalks" and are "buffeted / by a dark wind." The poem concludes suggesting that their greatest joy as well as their greatest strife is nature itself.
   c. The poem is an ode to the wind. Although the poem starts off comparing old age to birds, we see the poem calling upon the wind at the end in order to make sense of the world. The wind is ultimately in control in the poem—it is the driving force of the poem.
   d. This poem is about writing poetry. Old age signifies the poet, while the "small / cheeping birds" signifies the poet's words. The conclusion of the poem describes the wind as the creative process of writing a poem, and the "piping of plenty" is what the author gets in writing a fulfilling, lengthy poem.

4. What rhetorical device is shown in the very last line?
   a. Metaphor
   b. Simile
   c. Anaphora
   d. Alliteration

5. What time of year is the poem describing?
   a. Summer
   b. Fall
   c. Winter
   d. Spring

6. In the poem, what do the birds finally rest on?
   a. Seedhusks
   b. Weedstalks
   c. Trees
   d. The bare ground

The next article is for questions 7–11:

The Old Man and His Grandson

There was once a very old man, whose eyes had become dim, his ears dull of hearing, his knees trembled, and when he sat at table he could hardly hold the spoon, and spilt the broth upon the table-cloth or let it run out of his mouth. His son and his son's wife were disgusted at this, so the old grandfather at last had to sit in the corner behind the stove, and they gave him his food in an earthenware bowl, and not even enough of it. And he used to look towards the table with his eyes full of tears. Once, too, his trembling hands could not hold the bowl, and it fell to the ground and broke. The young wife scolded him, but he said nothing and only sighed. Then they brought him a wooden bowl for a few half-pence, out of which he had to eat.

They were once sitting thus when the little grandson of four years old began to gather together some bits of wood upon the ground. 'What are you doing there?' asked the father. 'I am making a little trough,' answered the child, 'for father and mother to eat out of when I am big.'

The man and his wife looked at each other for a while, and presently began to cry. Then they took the old grandfather to the table, and henceforth always let him eat with them, and likewise said nothing if he did spill a little of anything.

(Grimms' Fairy Tales, p. 111)

7. Which of the following most accurately represents the theme of the passage?
   a. Respect your elders
   b. Children will follow their parents' example
   c. You reap what you sow
   d. Loyalty will save your life

8. How is the content in this selection organized?
   a. Chronologically
   b. Problem and solution
   c. Compare and contrast
   d. Order of importance

9. Which character trait most accurately reflects the son and his wife in this story?
   a. Compassion
   b. Understanding
   c. Cruelty
   d. Impatience

10. Where does the story take place?
    a. In the countryside
    b. In America
    c. In a house
    d. In a forest

11. Why do the son and his wife decide to let the old man sit at the table?
    a. Because they felt sorry for him
    b. Because their son told them to
    c. Because the old man would not stop crying
    d. Because they saw their own actions in their son

In this excerpt from a novel set in nineteenth-century France, two friends, Albert de Morcerf and the Count of Monte Cristo, discuss Parisian social life. Read it and answer questions 12–18.

"Mademoiselle Eugénie is pretty—I think I remember that to be her name."

"Very pretty, or rather, very beautiful," replied Albert, "but of that style of beauty which I don't appreciate; I am an ungrateful fellow."

"Really," said Monte Cristo, lowering his voice, "you don't appear to me to be very enthusiastic on the subject of this marriage."

"Mademoiselle Danglars is too rich for me," replied Morcerf, "and that frightens me."

"Bah," exclaimed Monte Cristo, "that's a fine reason to give. Are you not rich yourself?"

"My father's income is about 50,000 francs per annum; and he will give me, perhaps, ten or twelve thousand when I marry."

"That, perhaps, might not be considered a large sum, in Paris especially," said the count; "but everything doesn't depend on wealth, and it's a fine thing to have a good name, and to occupy a high station in society. Your name is celebrated, your position magnificent; and then the Comte de Morcerf is a soldier, and it's pleasing to see the integrity of a Bayard united to the poverty of a Duguesclin; disinterestedness is the brightest ray in which a noble sword can shine. As for me, I consider the union with Mademoiselle Danglars a most suitable one; she will enrich you, and you will ennoble her."

Albert shook his head, and looked thoughtful. "There is still something else," said he.

"I confess," observed Monte Cristo, "that I have some difficulty in comprehending your objection to a young lady who is both rich and beautiful."

"Oh," said Morcerf, "this repugnance, if repugnance it may be called, isn't all on my side."

"Whence can it arise, then? for you told me your father desired the marriage."

"It's my mother who dissents; she has a clear and penetrating judgment, and doesn't smile on the proposed union. I cannot account for it, but she seems to entertain some prejudice against the Danglars."

"Ah," said the count, in a somewhat forced tone, "that may be easily explained; the Comtesse de Morcerf, who is aristocracy and refinement itself, doesn't relish the idea of being allied by your marriage with one of ignoble birth; that is natural enough."

12. The meaning of the word "repugnance" is closest to:
    a. Strong resemblance
    b. Strong dislike
    c. Extreme shyness
    d. Extreme dissimilarity

13. What can be inferred about Albert's family?
    a. Their finances are uncertain.
    b. Albert is the only son in his family.
    c. Their name is more respected than the Danglars'.
    d. Albert's mother and father both agree on their decisions.

14. What is Albert's attitude towards his impending marriage?
    a. Pragmatic
    b. Romantic
    c. Indifferent
    d. Apprehensive

15. What is the best description of the Count's relationship with Albert?
    a. He's like a strict parent, criticizing Albert's choices.
    b. He's like a wise uncle, giving practical advice to Albert.
    c. He's like a close friend, supporting all of Albert's opinions.
    d. He's like a suspicious investigator, asking many probing questions.

16. Which sentence is true of Albert's mother?
    a. She belongs to a noble family.
    b. She often makes poor choices.
    c. She is primarily occupied with money.
    d. She is unconcerned about her son's future.

17. Based on this passage, what is probably NOT true about French society in the 1800s?
    a. Children often received money from their parents.
    b. Marriages were sometimes arranged between families.
    c. The richest people in society were also the most respected.
    d. People were often expected to marry within their same social class.

18. Why is the Count puzzled by Albert's attitude toward his marriage?
    a. He seems reluctant to marry Eugénie, despite her wealth and beauty.
    b. He is marrying against his father's wishes, despite usually following his advice.
    c. He appears excited to marry someone he doesn't love, despite being a hopeless romantic.
    d. He expresses reverence towards Eugénie, despite being from a higher social class than her.

Questions 19-24 are based on the following passage.

Dana Gioia argues in his article that poetry is dying, now little more than a limited art form confined to academic and college settings. Of course, poetry remains healthy in the academic setting, but the idea of poetry being limited to this academic subculture is a stretch. New technology and social networking alone have contributed to poets and other writers' work being shared across the world. YouTube has emerged to be a major asset to poets, allowing live performances to be streamed to billions of users. Even now, poetry continues to grow and voice topics that are relevant to the culture of our time. Poetry is not in the spotlight as it may have been in earlier times, but it's still a relevant art form that continues to expand in scope and appeal.

Furthermore, Gioia's argument does not account for live performances of poetry. Not everyone has taken a poetry class or enrolled in university—but most everyone is online. The Internet is a perfect launching point to get all creative work out there. An example of this was the performance of Buddy Wakefield's *Hurling Crowbirds at Mockingbars*. Wakefield is a well-known poet who has published several collections of contemporary poetry. One of my favorite works by Wakefield is *Crowbirds*, specifically his performance at New York University in 2009. Although his reading was a campus event, views of his performance online number in the thousands. His poetry attracted people outside of the university setting.

Naturally, the poem's popularity can be attributed both to Wakefield's performance and the quality of his writing. *Crowbirds* touches on themes of core human concepts such as faith, personal loss, and growth. These are not ideas that only poets or students of literature understand, but all human beings: "You acted like I was hurling crowbirds at mockingbars / and abandoned me for not making sense. / Evidently, I don't experience things as rationally as you do" (Wakefield 15-17). Wakefield weaves together a complex description of the perplexed and hurt emotions of the speaker undergoing a separation from a romantic interest. The line "You acted like I was hurling crowbirds at mockingbars" conjures up an image of someone confused, seemingly out of their mind . . . or in the case of the speaker, passionately trying to grasp at a relationship that is fading. The speaker is looking back and finding the words that described how he wasn't making sense. This poem is particularly human and gripping in its message, but the entire effect of the poem is enhanced through the physical performance.

At its core, poetry is about addressing issues/ideas in the world. Part of this is also addressing the perspectives that are exiguously considered. Although the platform may look different, poetry continues to have a steady audience due to the emotional connection the poet shares with the audience.

19. Which one of the following best explains how the passage is organized?
    a. The author begins with a long definition of the main topic, and then proceeds to prove how that definition has changed over the course of modernity.
    b. The author presents a puzzling phenomenon and uses the rest of the passage to showcase personal experiences in order to explain it.
    c. The author contrasts two different viewpoints, then builds a case showing preference for one over the other.
    d. The passage is an analysis of another theory in which the author has no stake in.

20. The author of the passage would likely agree most with which of the following?
    a. Buddy Wakefield is a genius and is considered at the forefront of modern poetry.
    b. Poetry is not irrelevant; it is an art form that adapts to the changing time while containing its core elements.
    c. Spoken word is the zenith of poetic forms and the premier style of poetry in this decade.
    d. Poetry is on the verge of vanishing from our cultural consciousness.

21. Which one of the following words, if substituted for the word *exiguously* in the last paragraph, would LEAST change the meaning of the sentence?
    a. Indolently
    b. Inaudibly
    c. Interminably
    d. Infrequently

22. Which of the following is most closely analogous to the author's opinion of Buddy Wakefield's performance in relation to modern poetry?
    a. Someone's refusal to accept that the Higgs Boson will validate the Standard Model.
    b. An individual's belief that soccer will lose popularity within the next fifty years.
    c. A professor's opinion that poetry contains the language of the heart, while fiction contains the language of the mind.
    d. A student's insistence that psychoanalysis is a subset of modern psychology.

23. What is the primary purpose of the passage?
    a. To educate readers on the development of poetry and describe the historical implications of poetry in media.
    b. To disprove Dana Gioia's stance that poetry is becoming irrelevant and is only appreciated in academia.
    c. To inform readers of the brilliance of Buddy Wakefield and to introduce them to other poets that have influence in contemporary poetry.
    d. To prove that Gioia's article does have some truth to it and to shed light on its relevance to modern poetry.

24. What is the author's main reason for including the quote in the passage?
    a. The quote opens up opportunity to disprove Gioia's views.
    b. To demonstrate that people are still writing poetry even if the medium has changed in current times.
    c. To prove that poets still have an audience to write for even if the audience looks different than it did centuries ago.
    d. The quote illustrates the complex themes poets continue to address, which still draws listeners and appreciation.

Questions 25–29 are based upon the following passage:

This excerpt is adaptation from "The 'Hatchery' of the Sun-Fish"-- Scientific American, #711

I have thought that an example of the intelligence (instinct?) of a class of fish which has come under my observation during my excursions into the Adirondack region of New York State might possibly be of interest to your readers, especially as I am not aware that any one except myself has noticed it, or, at least, has given it publicity.

The female sun-fish (called, I believe, in England, the roach or bream) makes a "hatchery" for her eggs in this wise. Selecting a spot near the banks of the numerous lakes in which this region abounds, and where the water is about 4 inches deep, and still, she builds, with her tail and snout, a circular embankment 3 inches in height and 2 thick. The circle, which is as perfect a one as could be formed with mathematical instruments, is usually a foot and a half in diameter; and at one side of this circular wall an opening is left by the fish of just sufficient width to admit her body.

The mother sun-fish, having now built or provided her "hatchery," deposits her spawn within the circular inclosure, and mounts guard at the entrance until the fry are hatched out and are sufficiently large to take charge of themselves. As the embankment, moreover, is built up to the surface of the water, no enemy can very easily obtain an entrance within the inclosure from the top; while there being only one entrance, the fish is able, with comparative ease, to keep out all intruders.

I have, as I say, noticed this beautiful instinct of the sun-fish for the perpetuity of her species more particularly in the lakes of this region; but doubtless the same habit is common to these fish in other waters.

25. What is the purpose of this passage?
    a. To show the effects of fish hatcheries on the Adirondack region
    b. To persuade the audience to study Ichthyology (fish science)
    c. To depict the sequence of mating among sun-fish
    d. To enlighten the audience on the habits of sun-fish and their hatcheries

26. What does the word *wise* in this passage most closely mean?
    a. Knowledge
    b. Manner
    c. Shrewd
    d. Ignorance

27. What is the definition of the word *fry* as it appears in the following passage?
    The mother sun-fish, having now built or provided her "hatchery," deposits her spawn within the circular inclosure, and mounts guard at the entrance until the fry are hatched out and are sufficiently large to take charge of themselves.

    a. Fish at the stage of development where they are capable of feeding themselves.
    b. Fish eggs that have been fertilized.
    c. A place where larvae is kept out of danger from other predators.
    d. A dish where fish is placed in oil and fried until golden brown.

28. How is the circle that keeps the larvae of the sun-fish made?
    a. It is formed with mathematical instruments.
    b. The sun-fish builds it with her tail and snout.
    c. It is provided to her as a "hatchery" by Mother Nature.
    d. The sun-fish builds it with her larvae.

29. The author included the third paragraph in the following passage to achieve which of the following effects?
    a. To complicate the subject matter
    b. To express a bias
    c. To insert a counterargument
    d. To conclude a sequence and add a final detail

Questions 30-33 are based upon the following passage.

This excerpt is adaptation from Mineralogy --- Encyclopedia International, Grolier

Mineralogy is the science of minerals, which are the naturally occurring elements and compounds that make up the solid parts of the universe. Mineralogy is usually considered in terms of materials in the Earth, but meteorites provide samples of minerals from outside the Earth.

A mineral may be defined as a naturally occurring, homogeneous solid, inorganically formed, with a definite chemical composition and an ordered atomic arrangement. The qualification *naturally occurring* is essential because it is possible to reproduce most minerals in the laboratory. For example, evaporating a solution of sodium chloride produces crystal indistinguishable from those of the mineral halite, but such laboratory-produced crystals are not minerals.

A *homogeneous solid* is one consisting of a single kind of material that cannot be separated into simpler compounds by any physical method. The requirement that a mineral be solid eliminates gases and liquids from consideration. Thus ice is a mineral (a very common one, especially at high altitudes and latitudes) but water is not. Some mineralogists dispute this restriction and would consider both water and native mercury (also a liquid) as minerals.

The restriction of minerals to *inorganically formed* substances eliminates those homogenous solids produced by animals and plants. Thus the shell of an oyster and the pearl inside, though both consist of calcium carbonate indistinguishable chemically or physically from the mineral aragonite comma are not usually considered minerals.

The requirement of a *definite chemical composition* implies that a mineral is a chemical compound, and the composition of a chemical compound is readily expressed by a formula. Mineral formulas may be simple or complex, depending upon the number of elements present and the proportions in which they are combined.

Minerals are crystalline solids, and the presence of an *ordered atomic arrangement* is the criterion of the crystalline state. Under favorable conditions of formation the ordered atomic arrangement is expressed in the external crystal form. In fact, the presence of an ordered atomic arrangement and crystalline solids was deduced from the external regularity of crystals by a French mineralogist, Abbé R. Haüy, early in the 19th century.

30. According to the text, an object or substance must have all of the following criteria to be considered a mineral except for?
 a. Be naturally occurring
 b. Be a homogeneous solid
 c. Be organically formed
 d. Have a definite chemical composition

31. What is the definition of the word "homogeneous" as it appears in the following passage?

"A homogeneous solid is one consisting of a single kind of material that cannot be separated into simpler compounds by any physical method."
 a. Made of similar substances
 b. Differing in some areas
 c. Having a higher atomic mass
 d. Lacking necessary properties

32. The suffix -logy refers to?
 a. The properties of
 b. The chemical makeup of
 c. The study of
 d. The classification of

33. The author included the counterargument in the following passage to achieve which following effect?

The requirement that a mineral be solid eliminates gases and liquids from consideration. Thus, ice is a mineral (a very common one, especially at high altitudes and latitudes) but water is not. Some mineralogists dispute this restriction and would consider both water and native mercury (also a liquid) as minerals.
 a. To complicate the subject matter
 b. To express a bias
 c. To point to the fact that there are differing opinions in the field of mineralogy concerning the characteristics necessary to determine whether a substance or material is a mineral
 d. To create a new subsection of minerals

The following excerpt is from the article "The Lancashire Witches 1612–2012," by Robert Poole. Please read it and answer questions 34–39.

Four hundred years ago, in 1612, the north-west of England was the scene of England's biggest peacetime witch trial: the trial of the Lancashire witches. Twenty people, mostly from the Pendle area of Lancashire, were imprisoned in the castle as witches. Ten were hanged, one died in gaol, one was sentenced to stand in the pillory, and eight were acquitted. The 2012 anniversary sees a small flood of commemorative events, including works of fiction by Blake Morrison, Carol Ann Duffy, and Jeanette Winterson. How did this witch trial come about, and what accounts for its enduring fame?

We know so much about the Lancashire Witches because the trial was recorded in unique detail by the clerk of the court, Thomas Potts, who published his account soon afterwards as *The Wonderful Discovery of Witches in the County of Lancaster*. I have recently published a modern-English edition of this book, together with an essay piecing together what we know of the events of 1612. It has been a fascinating exercise, revealing how Potts carefully edited the evidence, and also how the case against the "witches" was constructed and manipulated to bring about a spectacular show trial. It all began in mid-March when a pedlar from Halifax named John Law had a frightening encounter with a poor young woman, Alizon Device, in a field near Colne. He refused her request for pins and there was a brief argument during which he was seized by a fit that left him with "his head … drawn awry, his eyes and face deformed, his speech not well to be understood; his thighs and legs stark lame." We can now recognize this as a stroke, perhaps triggered by the stressful encounter. Alizon Device was sent for and surprised all by confessing to the bewitching of John Law and then begged for forgiveness.

When Alizon Device was unable to cure the pedlar, the local magistrate, Roger Nowell was called in. Characterized by Thomas Potts as "God's justice" he was alert to instances of witchcraft, which were regarded by the Lancashire's puritan-inclined authorities as part of the cultural rubble of "popery"—Roman Catholicism—long overdue to be swept away at the end of the county's very slow protestant reformation. "With weeping tears" Alizon explained that she had been led astray by her grandmother, "old Demdike," well-known in the district for her knowledge of old Catholic prayers, charms, cures, magic, and curses. Nowell quickly interviewed Alizon's grandmother and mother, as well as Demdike's supposed rival, "old Chattox" and her daughter Anne. Their panicky attempts to explain themselves and shift the blame to others eventually only ended up incriminating them, and the four were sent to Lancaster gaol in early April to await trial at the summer assizes. The initial picture revealed was of a couple of poor, marginal local families in the forest of Pendle with a longstanding reputation for magical powers, which they had occasionally used at the request of their wealthier neighbours. There had been disputes but none of these were part of ordinary village life. Not until 1612 did any of this come to the attention of the authorities.

The net was widened still further at the end of April when Alizon's younger brother James and younger sister Jennet, only nine years old, came up between them with a story about a "great meeting of witches" at their grandmother's house, known as Malkin Tower. This meeting was presumably to discuss the plight of those arrested and the threat of further arrests, but according to the evidence extracted from the children by the magistrates, a plot was hatched to blow up Lancaster castle with gunpowder, kill the gaoler, and rescue the imprisoned witches. It was, in short, a conspiracy against royal authority to rival the gunpowder plot of 1605—something to be expected in a county known for its particularly strong underground Roman Catholic presence.

Those present at the meeting were mostly family members and neighbours, but they also included Alice Nutter, described by Potts as "a rich woman [who] had a great estate, and children of good hope: in the common opinion of the world, of good temper, free from envy or malice." Her part in the affair remains mysterious, but she seems to have had Catholic family connections, and may have been one herself, providing an added motive for her to be prosecuted.

34. What's the point of this passage, and why did the author write it?
    a. The author is documenting a historic witchcraft trial while uncovering/investigating the role of suspicion and anti-Catholicism in the events.
    b. The author seeks long-overdue reparations for the ancestors of those accused and executed for witchcraft in Lancashire.
    c. The author is educating the reader about actual occult practices of the 1600s.
    d. The author argues that the Lancashire witch trials were more brutal than the infamous Salem trials.

35. Which term best captures the meaning of the author's use of "enduring" in the first paragraph?
    a. Un-original
    b. Popular
    c. Wicked
    d. Circumstantial

36. What textual information is present within the passage that most lends itself to the author's credibility?
    a. His prose is consistent with the time.
    b. This is a reflective passage; the author doesn't need to establish credibility.
    c. The author cites specific quotes.
    d. The author has published a modern account of the case and has written on the subject before.

37. What might the following excerpt suggest about the trial or, at the very least, Thomas Potts' account of the trial(s)?
    "It has been a fascinating exercise, revealing how Potts carefully edited the evidence, and also how the case against the 'witches' was constructed and manipulated to bring about a spectacular show trial."

    a. The events were so grand that the public was allowed access to such a spectacular set of cases.
    b. Sections may have been exaggerated or stretched to create notoriety on an extraordinary case.
    c. Evidence was faked, making the trial a total farce.
    d. The trial was corrupt from the beginning.

38. Which statement best describes the political atmosphere of the 1600s that influenced the Alizon Device witch trial/case?
    a. Fear of witches was prevalent during this period.
    b. Magistrates were seeking ways to cement their power during this period of unrest.
    c. In a highly superstitious culture, the Protestant church and government were highly motivated to root out any potential sources that could undermine the current regime.
    d. Lancashire was originally a prominent area for pagan celebration, making the modern Protestants very weary of whispers of witchcraft and open to witch trials to resolve any potential threats to Christianity.

39. Which best describes the strongest "evidence" used in the case against Alizon and the witches?
    a. Knowledge of the occult and witchcraft
    b. "Spectral evidence"
    c. Popular rumors of witchcraft and Catholic association
    d. Self-incriminating speech

Questions 40–44 are based on the following passage.

George Washington emerged out of the American Revolution as an unlikely champion of liberty. On June 14, 1775, the Second Continental Congress created the Continental Army, and John Adams, serving in the Congress, nominated Washington to be its first commander. Washington fought under the British during the French and Indian War, and his experience and prestige proved instrumental to the American war effort. Washington provided invaluable leadership, training, and strategy during the Revolutionary War. He emerged from the war as the embodiment of liberty and freedom from tyranny.

After vanquishing the heavily favored British forces, Washington could have pronounced himself as the autocratic leader of the former colonies without any opposition, but he famously refused and returned to his Mount Vernon plantation. His restraint proved his commitment to the fledgling state's republicanism. Washington was later unanimously elected as the first American president. But it is Washington's farewell address that cemented his legacy as a visionary worthy of study.

In 1796, President Washington issued his farewell address by public letter. Washington enlisted his good friend, Alexander Hamilton, in drafting his most famous address. The letter expressed Washington's faith in the Constitution and rule of law. He encouraged his fellow Americans to put aside partisan differences and establish a national union. Washington warned Americans against meddling in foreign affairs and entering military alliances. Additionally, he stated his opposition to national political parties, which he considered partisan and counterproductive.

Americans would be wise to remember Washington's farewell, especially during presidential elections when politics hits a fever pitch. They might want to question the political institutions that were not planned by the Founding Fathers, such as the nomination process and political parties themselves.

40. Which of the following statements is based on the information in the passage above?
a. George Washington's background as a wealthy landholder directly led to his faith in equality, liberty, and democracy.
b. George Washington would have opposed America's involvement in the Second World War.
c. George Washington would not have been able to write as great a farewell address without the assistance of Alexander Hamilton.
d. George Washington would probably not approve of modern political parties.

41. What is the purpose of this passage?
a. To inform American voters about a Founding Father's sage advice on a contemporary issue and explain its applicability to modern times
b. To introduce George Washington to readers as a historical figure worthy of study
c. To note that George Washington was more than a famous military hero
d. To convince readers that George Washington is a hero of republicanism and liberty

42. What is the tone of the passage?
a. Informative
b. Excited
c. Bitter
d. Comic

43. What does the word *meddling* mean in paragraph 3?
    a. Supporting
    b. Speaking against
    c. Interfering
    d. Gathering

44. According to the passage, what did George Washington do when he was offered a role as leader of the former colonies?
    a. He refused the offer.
    b. He accepted the offer.
    c. He became angry at the offer.
    d. He accepted the offer then regretted it later.

The next question is based on the following passage.

> A famous children's author recently published a historical fiction novel under a pseudonym; however, it did not sell as many copies as her children's books. In her earlier years, she had majored in history and earned a graduate degree in Antebellum American History, which is the time frame of her new novel. Critics praised this newest work far more than the children's series that made her famous. In fact, her new novel was nominated for the prestigious Albert J. Beveridge Award, but still isn't selling like her children's books, which fly off the shelves because of her name alone.

45. Which one of the following statements might be accurately inferred based on the above passage?
    a. The famous children's author produced an inferior book under her pseudonym.
    b. The famous children's author is the foremost expert on Antebellum America.
    c. The famous children's author did not receive the bump in publicity for her historical novel that it would have received if it were written under her given name.
    d. People generally prefer to read children's series than historical fiction.

The next question is based on the following passage.

> In 2015, 28 countries, including Estonia, Portugal, Slovenia, and Latvia, scored significantly higher than the United States on standardized high school math tests. In the 1960s, the United States consistently ranked first in the world. Today, the United States spends more than $800 billion dollars on education, which exceeds the next highest country by more than $600 billion dollars. The United States also leads the world in spending per school-aged child by an enormous margin.

46. If these statements above are factual, which of the following statements must be correct?
    a. Outspending other countries on education has benefits beyond standardized math tests.
    b. The United States' education system is corrupt and broken.
    c. The standardized math tests are not representative of American academic prowess.
    d. Spending more money does not guarantee success on standardized math tests.

Math

1. Using the following diagram, what is the total circumference, rounding to the nearest decimal place?

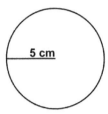

5 cm

    a. 25.0 cm
    b. 15.7 cm
    c. 78.5 cm
    d. 31.4 cm

2. What is the value of x in the following expression?
$$4x - 12 = -2x$$
    a. 2
    b. 3
    c. -6
    d. -2

3. What are the zeros of the function: $f(x) = x^3 + 4x^2 + 4x$?
    a. -2
    b. 0, -2
    c. 2
    d. 0, 2

4. What is the simplified quotient of the following expression?
$$\frac{5x^3}{3x^2y} \div \frac{25}{3y^9}$$
    a. $\frac{125x}{9y^{10}}$

    b. $\frac{x}{5y^8}$

    c. $\frac{5}{xy^8}$

    d. $\frac{xy^8}{5}$

5. What are the y-intercept(s) for $y = x^2 + 3x - 4$?
    a. $y = 1$
    b. $y = -4$
    c. $y = 3$
    d. $y = 4$

6. The area of a given rectangle is 24 centimeters. If the measure of each side is multiplied by 3, what is the area of the new figure?
    a. 48cm
    b. 72cm
    c. 216cm
    d. 13,824cm

7. Convert $\frac{5}{8}$ to a decimal.
    a. 0.62
    b. 1.05
    c. 0.63
    d. 1.60

8. How could the following equation be factored to find the zeros?
$$y = x^3 - 3x^2 - 4x$$
    a. $0 = x^2(x - 4), x = 0, 4$

    b. $0 = 3x(x + 1)(x + 4), x = 0, -1, -4$

    c. $0 = x(x + 1)(x + 6), x = 0, -1, -6$

    d. $0 = x(x + 1)(x - 4), x = 0, -1, 4$

9. In the figure below, what is the area of the shaded region?

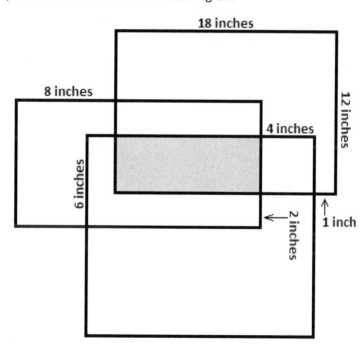

    a. 48 sq. inches
    b. 52 sq. inches
    c. 44 sq. inches
    d. 56 sq. inches

10. If the ordered pair $(-3, -4)$ is reflected over the x-axis, what's the new ordered pair?
    a. $(-3, -4)$
    b. $(3, -4)$
    c. $(3, 4)$
    d. $(-3, 4)$

11. What is the solution for the following equation?
$$\frac{x^2 + x - 30}{x - 5} = 11$$
    a. $x = -6$
    b. There is no solution.
    c. $x = 16$
    d. $x = 5$

12. Subtract and express in reduced form $\frac{23}{24} - \frac{1}{6}$.
    a. $\frac{22}{18}$
    b. $\frac{11}{9}$
    c. $\frac{19}{24}$
    d. $\frac{4}{5}$

13. Jessica buys 10 cans of paint. Red paint costs $1 per can, and blue paint costs $2 per can. In total, she spends $16. How many red cans did she buy?
    a. 2
    b. 3
    c. 4
    d. 6

14. Divide $702 \div 2.6$.
    a. 27
    b. 207
    c. 2.7
    d. 270

15. Multiply $13,114 \times 191$.
    a. 2,504,774
    b. 250,477
    c. 150,474
    d. 2,514,774

16. A rectangle was formed out of pipe cleaner. Its length was $\frac{1}{2}$ feet and its width was $\frac{11}{2}$ inches. What is its area in square inches?

    a. $\frac{11}{4}$ inch²

    b. $\frac{11}{2}$ inch²

    c. 22 inch²

    d. 33 inch²

17. The graph shows the position of a car over a 10-second time interval. Which of the following is the correct interpretation of the graph for the interval 1 to 3 seconds?

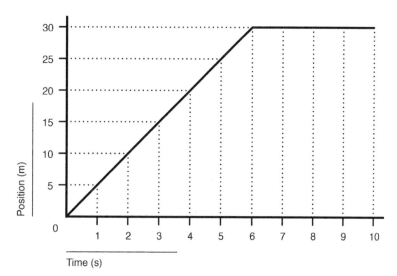

    a. The car remains in the same position.
    b. The car is traveling at a speed of 5m/s.
    c. The car is traveling up a hill.
    d. The car is traveling at 5mph.

18. Multiply and reduce $\frac{15}{23} \times \frac{54}{127}$.

    a. $\frac{810}{2,921}$

    b. $\frac{81}{292}$

    c. $\frac{69}{150}$

    d. $\frac{810}{2929}$

19. What is the sum of $(2.6 \times 10^5) + (1.3 \times 10^4)$?
    a. 3.38×10^5
    b. 2.73×10^5
    c. 1.3×10^4
    d. 2.47×10^4

20. What would the equation be for the following problem?

    3 times the sum of a number and 7 is greater than or equal to 32

    a. $3(7n) > 32$
    b. $3 \times n + 7 \geq 32$
    c. $3n + 21 > 32$
    d. $3(n + 7) \geq 32$

21. The table below shows tickets purchased during the week for entry to the local zoo. What is the mean of adult tickets sold for the week?

Day of the Week	Age	Tickets Sold
Monday	Adult	22
Monday	Child	30
Tuesday	Adult	16
Tuesday	Child	15
Wednesday	Adult	24
Wednesday	Child	23
Thursday	Adult	19
Thursday	Child	26
Friday	Adult	29
Friday	Child	38

    a. 24.2
    b. 21
    c. 22
    d. 26.4

22. Courtney leaves home and drives her truck 1,236 yards towards her destination. Her destination is 6,292 feet away from her home. How many more feet does she need to travel before she arrives?
    a. 2,284
    b. 3,708
    c. 5,056
    d. 2,584

23. A local candy store reports that of the 100 customers that bought suckers, 35 of them bought cherry. What is the probability of selecting 2 customers simultaneously at random that both purchased a cherry sucker?

    a. $\dfrac{119}{990}$

    b. $\dfrac{35}{100}$

    c. $\dfrac{49}{400}$

    d. $\dfrac{69}{99}$

24. Which inequality represents the number line below?

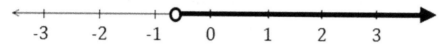

    a. $4x + 5 < 8$
    b. $-4x + 5 < 8$
    c. $-4x + 5 > 8$
    d. $4x - 5 > 8$

25. Which of the following formulas would correctly calculate the perimeter of a legal-sized piece of paper that is 14 inches long and $8\frac{1}{2}$ inches wide?

    a. $P = 14 + 8\frac{1}{2}$

    b. $P = 14 + 8\frac{1}{2} + 14 + 8\frac{1}{2}$

    c. $P = 14 \times 8\frac{1}{2}$

    d. $P = 14 \times \frac{17}{2}$

26. What is the least common multiple of 8 and 12?
    a. 36
    b. 24
    c. 48
    d. 32

27. What is the slope of the line that passes through the points $(10, -4)$ and $(-5, 8)$?

    a. $-\dfrac{5}{4}$

    b. $-\dfrac{4}{15}$

    c. $-\dfrac{4}{5}$

    d. $-\dfrac{12}{5}$

28. A grocery store is selling individual bottles of water, and each bottle contains 750 milliliters of water. If 12 bottles are purchased, how many liters did that customer take home?
    a. .9 liters
    b. 9 liters
    c. 900 liters
    d. 90 liters

29. What is the solution to the radical equation $\sqrt[3]{2x + 11} + 9 = 12$?
    a. -8
    b. 8
    c. 0
    d. 12

30. What is the solution to $(2 \times 20) \div (7 + 1) + (6 \times 0.01) + (4 \times 0.001)$?
    a. 5.064
    b. 5.64
    c. 5.0064
    d. 48.064

31. What is the solution to the following system of equations?
$$\begin{cases} x^2 + y = 4 \\ 2x + y = 1 \end{cases}$$

    a. $(-1, 3)$
    b. $(-1, 3), (3, -5)$
    c. $(3, -5)$
    d. $-1, 3$

32. What is the measurement of angle f in the following picture? Assume the lines are parallel.

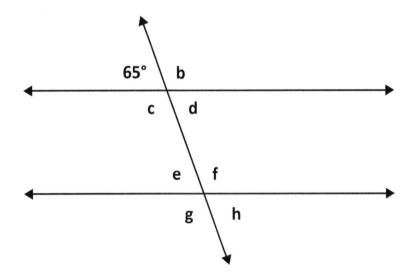

    a. 65 degrees
    b. 115 degrees
    c. 125 degrees
    d. 55 degrees

33. What is the product of the following expression?
$$(x + 2)(x^2 + 5x - 6)$$
a. $8x^2 + 4x - 12$
b. $x^2 + 6x - 4$
c. $x^3 + 7x^2 + 4x - 12$
d. $x^3 + 5x^2 - 4x - 12$

34. A piggy bank contains 12 dollars' worth of nickels. A nickel weighs 5 grams, and the empty piggy bank weighs 1050 grams. What is the total weight of the full piggy bank?
a. 1,110 grams
b. 1,200 grams
c. 2,250 grams
d. 2,200 grams

35. What is the value of x in the following triangle?

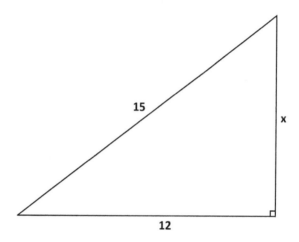

a. 19.2
b. 9
c. 3
d. 7.5

36. Last year, the New York City area received approximately $27\frac{3}{4}$ inches of snow. The Denver area received approximately 3 times as much snow as New York City. How much snow fell in Denver?
a. 60 inches

b. $27\frac{1}{4}$ inches

c. $9\frac{1}{4}$ inches

d. $83\frac{1}{4}$ inches

37. If $f(x) = x^2 - 3x + 17$, then what is $f(x + 1)$?
   a. $x^2 - 3x + 19$
   b. $x^2 - x + 15$
   c. $x^2 + 2x + 18$
   d. $x^2 - 3x + 14$

38. A group of marathon runners were surveyed to see how many miles they ran in a week to prepare for the upcoming race. Six runners were surveyed, and the median of the responses was 14 miles. Which of the following is a possible list of responses?
   a. 11, 12, 15, 16, 11, 13
   b. 13, 16, 11, 12, 15, 16
   c. 16, 16, 14, 13, 11, 12
   d. 12, 15, 16, 13, 16, 15

39. 250 students were asked their favorite flavor of ice cream. The results are presented in the pie chart below. What is the probability of choosing a student who likes vanilla or strawberry ice cream to receive a free ice cream cone?

Favorite Ice Cream Flavor

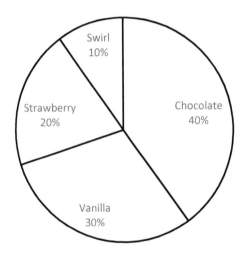

   a. $\frac{3}{10}$

   b. $\frac{2}{10}$

   c. $\frac{1}{2}$

   d. $\frac{1}{20}$

40. An employee receives 12 percent commission on every sale. What was the sale price of a stereo system if the commission received was $156?
   a. $1,350
   b. $1,300
   c. $1,200
   d. $1,400

41. The hospital has a nurse to patient ratio of 1:25. If there are a maximum of 325 patients admitted at a time, how many nurses are there?
   a. 13 nurses
   b. 25 nurses
   c. 325 nurses
   d. 12 nurses

42. A bucket can hold 11.4 liters of water. A kiddie pool needs 35 gallons of water to be full. How many times will the bucket need to be filled to fill the kiddie pool?
   a. 12
   b. 35
   c. 11
   d. 45

43. Solve for x: $\frac{2x}{5} - 1 = 59$.
   a. 60
   b. 145
   c. 150
   d. 115

44. A National Hockey League store in the state of Michigan advertises 50% off all items. Sales tax in Michigan is 6%. How much would a hat originally priced at $32.99 and a jersey originally priced at $64.99 cost during this sale? Round to the nearest penny.
   a. $97.98
   b. $103.86
   c. $51.93
   d. $48.99

45. Store brand coffee beans cost $1.23 per pound. A local coffee bean roaster charges $1.98 per $1\frac{1}{2}$ pounds. How much more would 5 pounds from the local roaster cost than 5 pounds of the store brand?
   a. $0.55
   b. $1.55
   c. $1.45
   d. $0.45

46. Paint Inc. charges $2000 for painting the first 1,800 feet of trim on a house and $1.00 per foot for each foot after. How much would it cost to paint a house with 3125 feet of trim?
   a. $3125
   b. $2000
   c. $5125
   d. $3325

47. A hospital has a bed to room ratio of 2: 1. If there are 145 rooms, how many beds are there?
    a. 145 beds
    b. 2 beds
    c. 90 beds
    d. 290 beds

48. Points L and M lie on line KN. The length of line KN is 30 units long, LN is 20 units long, and KM is 16 units long. How many units long is LM?

    a. 16
    b. 14
    c. 10
    d. 6

49. Angle y measures 48°. Angle x measures twice the value of angle y. What is the value of angle z?

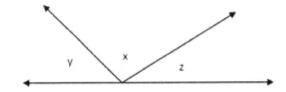

    a. 96°

    b. 36°

    c. 132°

    d. 42°

50. Evaluate $|4 - 9.6| + |12.3 - (-5)^3| =$
    a. 142.9
    b. 118.3
    c. 131.7
    d. 107.1

51. If $x = 2.6$ and $y = 5.3$, what is the value of $3.2xy - 4.1y$?
    a. -7.95
    b. 33.436
    c. 44.096
    d. 22.366

52. A data set is comprised of the following values: 30, 33, 33, 26, 27, 32, 33, 35, 29, 27. Which of the following has the greatest value?
    a. Mean
    b. Median
    c. Mode
    d. Range

53. Sam is twice as old as his sister, Lisa. Their oldest brother, Ray, will be 25 in three years. If Lisa is 13 years younger than Ray, how old is Sam?

54. What is the perimeter of the following figure rounded to the nearest tenth?

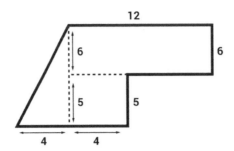

55. Solve the following:

$$\left(\sqrt{36} \times \sqrt{16}\right) - 3^2$$

56. What is the overall median of Dwayne's current scores: 78, 92, 83, 97?

57. The total perimeter of a rectangle is 36 cm. If the length of each side is 12 cm, what is the width?

Answer Explanations #2

Editing/Revising

1. A: Choice *A* is the correct answer. The *'s* in *Juliet's* suggests a contraction using the word *is*. Proper nouns cannot be part of a contraction, so *Juliet's* should be changed to *Juliet is*. The word *well* is an adverb that modifies known, and *known* is an adjective. They should be joined with a hyphen to describe the word, *tragedy*.

2. B: Choice *B* is correct because sentence 2 is punctuated incorrectly. This sentence is a complex sentence, or a sentence consisting of a dependent and independent clause. The phrase, *When air is inhaled*, is a clause because it starts with the subordinating conjunction, *when*. The phrase cannot stand alone as a complete sentence, so a comma is needed after the word *inhaled*.

3. C: Choice *C* is correct. These two sentences aim to provide a visual of how far a driver can travel while reading a text. Combining these sentences clarifies the idea that the amount of time it takes to drive the length of a football field is the minimum amount of time it takes to read or send a text while driving.

4. D: Choice *D* is the most concise way to convey the idea presented in the sentence. Choice *A* is incorrect because the revision only eliminates some of the wordiness in the sentence. Choice *B* is incorrect because it eliminates too much information, such as Mr. Wilkinson's name and the fact that he was late to work. Choice *C* is also incorrect because it leaves out some important information, such as the fact that Mr. Wilkinson was late to work, and the class was being rewarded for good test grades.

5. A: Choice *A* is correct because the word *Flag* does not need to be capitalized, and commas are needed to separate three or more words in a list, such as *unity, strength, and courage*.

6. B: Choice *B* combines the two sentences while still including the correct information. Choice *A* leaves out the point that the density of a penguin's bones helps it to float and dive. Choice *C* alludes to the idea that birds just can't fly because they are fat. Choice *D* does not pertain to the information in either of the original two sentences.

7. C: Choice *C* is the correct answer. In sentence 3, the word *Libertys* is plural, indicating more than one Liberty. However, adding an apostrophe changes the word from a plural to a noun showing ownership, her crown. The word *weighs* is a singular verb in this sentence but needs to be a plural verb. Plural subjects should be followed by plural verbs and vice versa. So, since the subject of the sentence, *rays*, is plural, the verb *weighs* needs to be plural, or *weigh*.

8. B: Choice *B* is the correct answer because it states a general fact about Japanese food and serves as a main idea for the paragraph. Although Choices *A, C,* and *D* are true facts about Japanese cuisine, they are details that would further support a main idea about Japanese food.

9. C: Choice *C* is the correct answer. Sentence 10 should be the next sentence after sentence 8. It introduces the town of Hakone and provides the framework for the specific information about the town presented in sentences 9 through 12.

10. D: Choice *D* is the correct answer. This sentence is more concise and still incorporates all of the information trying to be conveyed in the original sentence. Choices *A, B*, and *C* either mix up the information in the original sentence or give incorrect facts.

11. B: Choice *B* is the correct answer. In this sentence, the words *or bullet train* serve as an appositive, or extra information provided in the sentence. These words are added to the sentence to define *Shinkansen*. However, the sentence would be correct even if these words hadn't been added, so this phrase should be set apart with commas. A correct version of the sentence would look like this: [8]*People from all over Japan ride the Shinkansen, or bullet train, to Osaka to watch the events.*

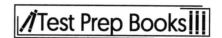

Reading Comprehension

1. C: The poem uses imagery. William Carlos Williams is considered a Modernist poet who relied heavily on imagery to bring poetry to life. Poets like William Carlos Williams, Ezra Pound, and Marianne Moore wrote imagist poems typical of their time period. The poem is not metrical, but written in free verse, so Choice *A* is incorrect. Anaphora, Choice *B*, is a repetition of words at the beginning of a succession of lines. Synecdoche, Choice *D*, refers to a part of something that represents a whole.

2. B: This poem comes from the Modernist period. Modernism is a literary movement featuring writers like William Carlos Williams, Ezra Pound, T.S. Eliot, Marianne Moore, and Wallace Stevens. The poetry is a reaction to traditional metrical poetry and its outdated language. Modernists rely on heavy imagery and sometimes short, terse stanzas to create the impact of disjointedness with the self and language.

3. A: The poem acts as an extended metaphor of old age. Its juxtaposed imagery suggests the frailty of life ("Gaining and failing / they are buffeted / by a dark wind") alongside the fullness of life and its "piping of plenty." The other choices may have some aspects that are true of an effective analysis; however, the most important thing to recognize here is that the poem starts off very clearly as an extended metaphor.

4. D: The rhetorical device used in the last line is alliteration: "piping of plenty." Alliteration is where the same sound is used for an auditory effect. Metaphor and simile compare two things to each other, so these are incorrect. Anaphora is when two or more lines use the same beginning repetition.

5. C: The poem is describing winter. There is "dark wind," "bare trees," "snow," and a "snow glaze," so we can infer that this is the coldest season of the year.

6. B: The birds finally rest on weedstalks. We see this in the inverted line, "On harsh weedstalks / the flock has rested."

7. B: *A* is incorrect because it does not fit with the primary purpose of this passage, which is to tell a story of how a child plans to treat his parents when he sees the way they treat his grandfather. It is trying to remind readers to treat others with respect because that is how one wants to be treated, and that this does not apply only to elderly people. Choice *B* fits most appropriately with the primary purpose, since the son and wife see that they will be treated unfairly because they witness that their child plans to do it to them when they are older. To "reap what you sow" means that there are repercussions for every action. This may seem like the correct answer; however, the parents do not actually have to eat out of a trough later in life. They don't actually experience any repercussions. Even though it may be argued that the boy is being loyal to his grandfather, this does not fit with the primary purpose. The boy also never mentions that his actions are because he cares for his grandfather; rather, he simply mirrors the behaviors of his parents.

8. A: *A* is correct because it follows a series of events that happen in order, one right after the other. First the grandfather spills his food, then his son puts him in a corner, then the child makes a trough for his parents to eat out of when he's older, and finally the parents welcome the old man back to the table. Choice *B* is incorrect as even though it could be argued that the way they treat the old man is a problem, there really isn't a solution to the problem, even though they stop treating him badly. Also, problem and solution styles generally do not follow a chronological timeline. Choice *C* is incorrect because events in the passage are not compared and contrasted; this is not a primary organizational structure of the passage. Choice *D* is incorrect because there is no language to indicate that one person or event is more important than the other.

9. C: Although they do show him compassion in the end, it is not because they feel compassionate for him, but instead, it is because they recognize that their son plans to treat them the way they are treating the old man when they are older. So, they treat the old man the way they would want to be treated. Understanding is not the overall attitude they feel toward the old man, and it is only in realizing the cruelty of their behavior that they understand how they have been treating him. Choice *C* is correct because it condenses the actions of the son and his wife into a single word. Refusing to let the old man sit at the table when he clearly needs help and looks at the table with tear-filled eyes is a cruel thing to do. Choice *D* may be tempting to pick as they *are* impatient with him, but it's not the best answer. People can be impatient without being cruel.

10. C: Choice *A* is incorrect as there is no descriptive language to indicate that they are in the countryside. *B* is incorrect because the passage has no language or descriptions to indicate they are in America. Choice *C* is correct because the setting contains elements of a house: a table, a stove, and a corner. Choice *D* may be tempting as there is mention of "bits of wood upon the ground," but as there are no other elements of a forest in the story, this is not the correct answer.

11. D: They allow the old man to sit at the table because their son starts to make them a trough, so their motivation in letting him eat at the table is not because they feel sorry for him, but because they don't want their son to treat them that way when they are old. This makes Choice *A* incorrect. Their son did not tell them to let the old man sit at the table, so Choice *B* is incorrect. In the story, it mentions that even after the old man has eyes full of tears, the wife gave him a cheap wooden bowl to eat out of, so clearly his crying did not make them stop treating him badly, making Choice *C* incorrect. Choice *D* is correct because the parents let the old man sit at the table as a result of the boy mimicking their behavior.

12. B: Strong dislike. This vocabulary question can be answered using context clues and common sense. Based on the rest of the conversation, the reader can gather that Albert isn't looking forward to his marriage. As the Count notes that "you don't appear to me to be very enthusiastic on the subject of this marriage," and also remarks on Albert's "objection to a young lady who is both rich and beautiful," readers can guess Albert's feelings. The answer choice that most closely matches "objection" and "not . . . very enthusiastic" is *B*, "strong dislike."

13. C: Their name is more respected than the Danglars'. This inference question can be answered by eliminating incorrect answers. Choice *A* is tempting, considering that Albert mentions money as a concern in his marriage. However, although he may not be as rich as his fiancée, his father still has a stable income of 50,000 francs a year. Choice *B* isn't mentioned at all in the passage, so it's impossible to make an inference. Finally, Choice *D* is clearly false because Albert's father arranged his marriage, but his mother doesn't approve of it. Evidence for Choice *C* can be found in the Count's comparison of

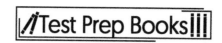

Albert and Eugénie: "she will enrich you, and you will ennoble her." In other words, the Danglars are wealthier but the Morcef family has a more noble background.

14. D: Apprehensive. As in question 7, there are many clues in the passage that indicate Albert's attitude towards his marriage—far from enthusiastic, he has many reservations. This question requires test takers to understand the vocabulary in the answer choices. "Pragmatic" is closest in meaning to "realistic," and "indifferent" means "uninterested." The only word related to feeling worried, uncertain, or unfavorable about the future is "apprehensive."

15. B: He is like a wise uncle, giving practical advice to Albert. Choice *A* is incorrect because the Count's tone is friendly and conversational. Choice *C* is also incorrect because the Count questions why Albert doesn't want to marry a young, beautiful, and rich girl. While the Count asks many questions, he isn't particularly "probing" or "suspicious"—instead, he's asking to find out more about Albert's situation and then give him advice about marriage.

16. A: She belongs to a noble family. Though Albert's mother doesn't appear in the scene, there's more than enough information to answer this question. More than once is his family's noble background mentioned (not to mention that Albert's mother is the Comtess de Morcef, a noble title). The other answer choices can be eliminated—she is obviously deeply concerned about her son's future; money isn't her highest priority because otherwise she would favor a marriage with the wealthy Danglars; and Albert describes her "clear and penetrating judgment," meaning she makes good decisions.

17. C: The richest people in society were also the most respected. The Danglars family is wealthier but the Morcef family has a more aristocratic name, which gives them a higher social standing. Evidence for the other answer choices can be found throughout the passage: Albert mentioned receiving money from his father's fortune after his marriage; Albert's father has arranged this marriage for him; and the Count speculates that Albert's mother disapproves of this marriage because Eugénie isn't from a noble background like the Morcef family, implying that she would prefer a match with a girl from aristocratic society.

18. A: He seems reluctant to marry Eugénie, despite her wealth and beauty. This is a reading comprehension question, and the answer can be found in the following lines: "'I confess,' observed Monte Cristo, 'that I have some difficulty in comprehending your objection to a young lady who is both rich and beautiful.'" Choice *B* is the opposite (Albert's father is the one who insists on the marriage), Choice *C* incorrectly represents Albert's eagerness to marry, and Choice *D* describes a more positive attitude than Albert actually feels ("repugnance").

19. C: The author contrasts two different viewpoints, then builds a case showing preference for one over the other. Choice *A* is incorrect because the introduction does not contain an impartial definition, but rather, an opinion. Choice *B* is incorrect. There is no puzzling phenomenon given, as the author doesn't mention any peculiar cause or effect that is in question regarding poetry. Choice *D* does contain another's viewpoint at the beginning of the passage; however, to say that the author has no stake in this argument is incorrect; the author uses personal experiences to build their case.

20. B: Choice *B* accurately describes the author's argument in the text: that poetry is not irrelevant. While the author does praise, and even value, Buddy Wakefield as a poet, the author never heralds him as a genius. Eliminate Choice *A*, as it is an exaggeration. Not only is Choice *C* an exaggerated statement, but the author never mentions spoken word poetry in the text. Choice *D* is wrong because this statement contradicts the writer's argument.

21. D: *Exiguously* means not occurring often, or occurring rarely, so Choice *D* would LEAST change the meaning of the sentence. Choice *A*, *indolently*, means unhurriedly, or slow, and does not fit the context of the sentence. Choice *B*, *inaudibly*, means quietly or silently. Choice *C*, *interminably*, means endlessly, or all the time, and is the opposite of the word *exiguously*.

22. D: A student's insistence that psychoanalysis is a subset of modern psychology is the most analogous option. The author of the passage tries to insist that performance poetry is a subset of modern poetry, and therefore, tries to prove that modern poetry is not "dying," but thriving on social media for the masses. Choice *A* is incorrect, as the author is not refusing any kind of validation. Choice *B* is incorrect; the author's insistence is that poetry will *not* lose popularity. Choice *C* mimics the topic but compares two different genres, while the author does no comparison in this passage.

23. B: The author's purpose is to disprove Gioia's article claiming that poetry is a dying art form that only survives in academic settings. In order to prove his argument, the author educates the reader about new developments in poetry (Choice *A*) and describes the brilliance of a specific modern poet (Choice *C*), but these serve as examples of a growing poetry trend that counters Gioia's argument. Choice *D* is incorrect because it contradicts the author's argument.

24. D: This question is difficult because the choices offer real reasons as to why the author includes the quote. However, the question specifically asks for the *main reason* for including the quote. The quote from a recently written poem shows that people are indeed writing, publishing, and performing poetry (Choice *B*). The quote also shows that people are still listening to poetry (Choice *C*). These things are true, and by their nature, serve to disprove Gioia's views (Choice *A*), which is the author's goal. However, Choice *D* is the most direct reason for including the quote, because the article analyzes the quote for its "complex themes" that "draws listeners and appreciation" right after it's given.

25. D: To enlighten the audience on the habits of sun-fish and their hatcheries. Choice *A* is incorrect because although the Adirondack region is mentioned in the text, there is no cause or effect relationships between the region and fish hatcheries depicted here. Choice *B* is incorrect because the text does not have an agenda, but rather is meant to inform the audience. Finally, Choice *C* is incorrect because the text says nothing of how sun-fish mate.

26. B: The word *wise* in this passage most closely means *manner*. Choices *A* and *C* are synonyms of *wise*; however, they are not relevant in the context of the text. Choice *D*, *ignorance*, is opposite of the word *wise*, and is therefore incorrect.

27. A: Fish at the stage of development where they are capable of feeding themselves. Even if the word *fry* isn't immediately known to the reader, the context gives a hint when it says "until the fry are hatched out and are sufficiently large to take charge of themselves."

28. B: The sun-fish builds it with her tail and snout. The text explains this in the second paragraph: "she builds, with her tail and snout, a circular embankment 3 inches in height and 2 thick." Choice *A* is used in the text as a simile.

29. D: To conclude a sequence and add a final detail. The concluding sequence is expressed in the phrase "[t]he mother sun-fish, having now built or provided her 'hatchery.'" The final detail is the way in which the sun-fish guards the "inclosure." Choices *A, B,* and *C* are incorrect.

30. C: The text mentions all of the listed properties of minerals except the instance of minerals being organically formed. Objects or substances must be naturally occurring, must be a homogeneous solid, and must have a definite chemical composition in order to be considered a mineral.

31. A: Choice *A* is the correct answer because the prefix "homo" means same. Choice *B* is incorrect because "differing in some areas" would be linked to the root word "hetero," meaning "different" or "other."

32: C: Choice *C* is the correct answer because *-logy* refers to the study of a particular subject matter.

33: C: Choice *C* is the correct answer because the counterargument is necessary to point to the fact that researchers don't always agree with findings. Choices *A* and *B* are incorrect because the counterargument isn't overcomplicated or expressing bias, but simply stating an objective dispute. Choice *D* is incorrect because the counterargument is not used to persuade readers to create a new subsection of minerals.

34. A: Choice *D* can be eliminated because the Salem witch trials aren't even mentioned. While sympathetic to the plight of the accused, the author doesn't demand or urge the reader to demand reparations to the descendants; therefore, Choice *B* can also be ruled out. It's clear that the author's main goal is to educate the reader and shed light on the facts and hidden details behind the case. However, his focus isn't on the occult, but the specific Lancashire case itself. He goes into detail about suspects' histories and ties to Catholicism, revealing how the fears of the English people at the time sealed the fate of the accused witches. Choice *A* is correct.

35. B: It's important to note that these terms may not be an exact analog for *enduring*. However, through knowledge of the definition of *enduring*, as well as the context in which it's used, an appropriate synonym can be found. Plugging "circumstantial" into the passage in place of "enduring" doesn't make sense. Nor does "un-original," this particular case of witchcraft, stand out in history. "Wicked" is very descriptive, but this is an attribute applied to people, not events; therefore, this is an inappropriate choice as well. *Enduring* literally means long lasting, referring to the continued interest in this particular case of witchcraft. Therefore, it's a popular topic of 1600s witch trials, making "popular," Choice *B*, the best choice.

36. D: Choices A and B are irrelevant. The use of quotes lends credibility to the author. However, the presence of quotes alone doesn't necessarily mean that the author has a qualified perspective. What establishes the writer as a reliable voice is that the author's previous writing on the subject has been published before. This qualification greatly establishes the author's credentials as a historical writer, making Choice D the correct answer.

37. B: Choice *A* is incorrect, clearly taking the statement somewhat literally. The remaining three choices appear somewhat interconnected, and though they may be proven at some point later in the article, the focus must remain on the given excerpt. It's very possible that evidence was tampered with or even falsified, but this statement doesn't refer to this. While the author alludes that there may have been evidence tampering and potentially corruption, what the writer is directly saying is that the documentation of the court indicates an elaborate trial. It's clear that exaggerations may have taken place both during the case and in the written account. The reasoning behind this was to gain the attention of the people and even the crown. Choice *B* is the best answer because it not only aligns with the above statement, but ultimately encompasses the potentiality of Choices C and D as well.

38. C: Several of these answers could have contributed to the fear and political motivations around the Lancashire witch trials. What this answer's looking for is very specific: political motivations and issues that played a major role in the case. Choice C clearly outlines the public fears of the time. It also describes how the government can use this fear to weed out and eliminate traces of Catholicism (and witchcraft too). Catholicism and witchcraft were seen as dangerous and undermining to English Protestantism and governance. Choice D can be eliminated; while this information may have some truth and is certainly consistent with the general fear of witchcraft, the details about Lancashire's ancient history aren't mentioned in the text. Choice A is true but not necessarily political in nature. Choice B is very promising, though not outright mentioned.

39. D: The best evidence comes from Alizon herself. The text mentions that she confessed to bewitching John Law, thinking that she did him harm. From here she names her grandmother, who she believes corrupted her. Choice B can be ruled out; spectral evidence isn't mentioned. The case draws on knowledge of superstition of witchcraft, but this in itself can't be considered evidence, so Choice A is incorrect. Choice C isn't evidence in a modern sense; rumors have no weight in court and therefore are not evidence. While this is used as evidence to some degree, this still isn't the best evidence against Alizon and the witches.

40. D: Although Washington is from a wealthy background, the passage does not say that his wealth led to his republican ideals, so Choice A is not supported. Choice B also does not follow from the passage. Washington's warning against meddling in foreign affairs does not mean that he would oppose wars of every kind, so Choice B is incorrect. Choice C is also unjustified since the author does not indicate that Alexander Hamilton's assistance was absolutely necessary. Choice D is correct because the farewell address clearly opposes political parties and partisanship. The author then notes that presidential elections often hit a fever pitch of partisanship. Thus, it is follows that George Washington would not approve of modern political parties and their involvement in presidential elections.

41. A: The author finishes the passage by applying Washington's farewell address to modern politics, so the purpose probably includes this application. Choice B is incorrect because George Washington is already a well-established historical figure; furthermore, the passage does not seek to introduce him. Choice C is incorrect because the author is not fighting a common perception that Washington was merely a military hero. Choice D is incorrect because the author is not convincing readers. Persuasion does not correspond to the passage. Choice A states the primary purpose.

42. A: The tone in this passage is informative. Choice B, excited, is incorrect, because there are not many word choices used that would indicate excitement from the author. Choice C, bitter, is incorrect. Although the author does make a suggestion in the last paragraph to Americans, the statement is not necessarily bitter, but based on the preceding information. Choice D, comic, is incorrect, as the author does not try to make the audience laugh, nor do they make light of the situation in any way.

43. C: Interfering. Meddling means to interfere in something. Choice A is incorrect. One helpful thing would be to use the word in the sentence: "Washington warned Americans against 'supporting' in foreign affairs" does not make that much sense, so we can mark it off. Choice B, speaking against, is incorrect. This phrase would make sense in the sentence, but it goes against the meaning that is intended. George Washington warned against interference in foreign affairs, not speaking *against* foreign affairs. Finally, gathering is also incorrect, because "gathering in foreign affairs" does not sound quite right. Interfering is the best choice for this question.

44. A: When Washington was offered a role as leader of the former colonies, he refused the offer. This is explained in the first sentence of the second paragraph. All of the other answer choices are incorrect and not mentioned in the passage.

45. C: We are looking for an inference—a conclusion that is reached on the basis of evidence and reasoning—from the passage that will likely explain why the famous children's author did not achieve her usual success with the new genre (despite the book's acclaim). Choice *A* is wrong because the statement is false according to the passage. Choice *B* is wrong because, although the passage says the author has a graduate degree on the subject, it would be an unrealistic leap to infer that she is the foremost expert on Antebellum America. Choice *D* is wrong because there is nothing in the passage to lead us to infer that people generally prefer a children's series to historical fiction. In contrast, Choice *C* can be logically inferred since the passage speaks of the great success of the children's series and the declaration that the fame of the author's name causes the children's books to "fly off the shelves." Thus, she did not receive any bump from her name since she published the historical novel under a pseudonym, and Choice *C* is correct.

46. D: Outspending other countries on education could have other benefits, but there is no reference to this in the passage, so Choice *A* is incorrect. Choice *B* is incorrect because the author does not mention corruption. Choice *C* is incorrect because there is nothing in the passage stating that the tests are not genuinely representative. Choice *D* is accurate because spending more money has not brought success. The United States already spends the most money, and the country is not excelling on these tests. Choice *D* is the correct answer.

![Test Prep Books logo]

Math

1. D: To calculate the circumference of a circle, use the formula $2\pi r$, where r equals the radius, or half of the diameter, of the circle and $\pi = 3.14 \ldots$. Substitute the given information, $2\pi 5 = 31.4 \ldots$, which is Choice *D*.

2. A: To solve for x the steps are as follows:

$$4x - 12$$

$$-2x, 6x - 12 = 0$$

$$6x - 12$$

$$x = 2$$

3. B: There are two zeros for the given function. They are $x = 0, -2$. The zeros can be found a number of ways, but this particular equation can be factored into:

$$f(x) = x(x^2 + 4x + 4)$$

$$x(x + 2)(x + 2)$$

By setting each factor equal to zero and solving for x, there are two solutions. On a graph, these zeros can be seen where the line crosses the x-axis.

4. D: Dividing rational expressions follows the same rule as dividing fractions. The division is changed to multiplication, and the reciprocal is found in the second fraction. This turns the expression into:

$$\frac{5x^3}{3x^2} \times \frac{3y^9}{25}$$

Multiplying across and simplifying, the final expression is:

$$\frac{xy^8}{5}$$

5. B: The y-intercept of an equation is found where the x-value is zero. Plugging zero into the equation for x, the first two terms cancel out, leaving -4.

6. C: 216cm. Because area is a two-dimensional measurement, the dimensions are multiplied by a scale that is squared to determine the scale of the corresponding areas. The dimensions of the rectangle are multiplied by a scale of 3. Therefore, the area is multiplied by a scale of 3^2 (which is equal to 9):

$$24 \ cm \times 9 = 216 \ cm$$

7. C: 0.63

Divide 5 by 8, which results in 0.625. This rounds up to 0.63.

8. D: Finding the zeros for a function by factoring is done by setting the equation equal to zero, then completely factoring. Since there was a common x for each term in the provided equation, that is factored out first.

Then the quadratic that is left can be factored into two binomials:

$$(x + 1)(x - 4)$$

Setting each factor equation equal to zero and solving for x yields three zeros.

9. B: This can be determined by finding the length and width of the shaded region. The length can be found using the length of the top rectangle, which is 18 inches, then subtracting the extra length of 4 inches and 1 inch. This means the length of the shaded region is 13 inches. Next, the width can be determined using the 6 inch measurement and subtracting the 2 inch measurement. This means that the width is 4 inches. Thus, the area is:

$$13 \times 4 = 52 \ sq. \ in.$$

10. D: When an ordered pair is reflected over an axis, the sign of one of the coordinates must change. When it's reflected over the x-axis, the sign of the x coordinate must change. The y value remains the same. Therefore, the new ordered pair is $(-3, 4)$.

11. B: The equation can be solved by factoring the numerator into $(x + 6)(x - 5)$. Since that same factor $(x - 5)$ exists on top and bottom, that factor cancels. This leaves the equation $x + 6 = 11$. Solving the equation gives the answer $x = 5$. When this value is plugged into the equation, it yields a zero in the denominator of the fraction. Since this is undefined, there is no solution.

12. C: $\frac{19}{24}$

Set up the problem and find a common denominator for both fractions.

$$\frac{23}{24} - \frac{1}{6}$$

Multiply each fraction across by 1 to convert to a common denominator.

$$\frac{23}{24} \times \frac{1}{1} - \frac{1}{6} \times \frac{4}{4}$$

Once over the same denominator, subtract across the top.

$$\frac{23 - 4}{24} = \frac{19}{24}$$

13. C: Let r be the number of red cans and b be the number of blue cans. One equation is:

$$r + b = 10$$

The total price is $16, and the prices for each can means:

$$1r + 2b = 16$$

Multiplying the first equation on both sides by -1 results in:

$$-r - b = -10$$

Add this equation to the second equation, leaving $b = 6$. So, she bought 6 *blue* cans. From the first equation, this means $r = 4$; thus, she bought 4 *red* cans.

14. D: 270

Set up the division problem.

$$2.6\overline{)702}$$

Move the decimal over one place to the right in both numbers.

$$26\overline{)7020}$$

26 does not go into 7 but does go into 70 so start there.

$$\begin{array}{r} 270 \\ 26\overline{)7020} \\ -52 \\ \hline 182 \\ -182 \\ \hline 0 \end{array}$$

The result is 270

15. A: 2,504,774

Line up the numbers (the number with the most digits on top) to multiply. Begin with the right column on top and the right column on bottom.

Move one column left on top and multiply by the far right column on the bottom. Remember to add the carry over after you multiply. Continue that pattern for each of the numbers on the top row.

Starting on the far right column on top repeat this pattern for the next number left on the bottom. Write the answers below the first line of answers; remember to begin with a zero placeholder. Continue for each number in the top row.

Starting on the far right column on top, repeat this pattern for the next number left on the bottom. Write the answers below the first line of answers. Remember to begin with zero placeholders.

Once completed, ensure the answer rows are lined up correctly, then add.

16. D: Area = length x width. The answer must be in square inches, so all values must be converted to inches. $\frac{1}{2}$ ft is equal to 6 inches. Therefore, the area of the rectangle is equal to:

$$6 \times \frac{11}{2}$$

$$\frac{66}{2}$$

33 square inches

on

17. B: The car is traveling at a speed of five meters per second. On the interval from one to three seconds, the position changes by fifteen meters. By making this change in position over time into a rate, the speed becomes ten meters in two seconds or five meters in one second.

18. A: $\frac{810}{2921}$

Line up the fractions.

$$\frac{15}{23} \times \frac{54}{127}$$

Multiply across the top and across the bottom.

$$\frac{15 \times 54}{23 \times 127} = \frac{810}{2921}$$

19. B: The exponent of the ten must be the same before any operations are performed on the numbers. So, $(2.6 \times 10^5) + (1.3 \times 10^4)$ cannot be added until one of the exponents on the ten is changed. The 1.3×10^4 can be changed to 0.13×10^5, then the 2.6 and 0.13 can be added. The answer is 2.73×10^5.

20. D: 3 times the sum of a number and 7 is greater than or equal to 32 can be translated into equation form utilizing mathematical operators and numbers.

21. C: To find the mean, or average, of a set of values, add the values together and then divide by the total number of values. Each day of the week has an adult ticket amount sold that must be added together. The equation is as follows:

$$\frac{22 + 16 + 24 + 19 + 29}{5} = 22$$

22. D: First, convert the distance that Courtney already drove to feet. Because there are three feet per yard, her distance traveled thus far in yards must be multiplied by 3:

$$1,236 \times 3 = 3,708 \text{ feet}$$

If the total distance to travel is 6,292 feet, there is $6292 - 3708 = 2,584$ feet left to travel.

23. A: The probability of choosing two customers simultaneously is the same as choosing one and then choosing a second without putting the first back into the pool of customers. This means that the probability of choosing a customer who bought cherry is $\frac{35}{100}$. Then without placing them back in the pool, it would be $\frac{34}{99}$.

So, the probability of choosing 2 customers simultaneously that both bought cherry would be:

$$\frac{35}{100} \times \frac{34}{99}$$

$$\frac{1,190}{9,900}$$

$$\frac{119}{990}$$

24. B: The number line shows:

$$x > -\frac{3}{4}$$

Each inequality must be solved for x to determine if it matches the number line. Choice A of $4x + 5 < 8$ results in $x < -\frac{3}{4}$, which is incorrect. Choice C of $-4x + 5 > 8$ yields $x < -\frac{3}{4}$, which is also incorrect. Choice D of $4x - 5 > 8$ results in $x > \frac{13}{4}$, which is not correct. Choice B, $-4x + 5 < 8$ is the only choice that results in the correct answer of:

$$x > -\frac{3}{4}$$

25. B: The perimeter of a rectangle is the sum of all four sides. Therefore, the answer is:

$$P = 14 + 8\frac{1}{2} + 14 + 8\frac{1}{2}$$

$$14 + 14 + 8 + \frac{1}{2} + 8 + \frac{1}{2} = 45 \text{ square inches.}$$

26. B: The least common multiple is the smallest number that is a multiple of two numbers. The first few multiples of 8 are 8, 16, 24, 32, 40, and 48. The first few multiples of 12 are 12, 24, 36, 48, and 60. Both 24 and 48 are common multiples of 8 and 12, but 24 is the least common multiple.

27. C: The equation used to find the slope of a line when given two points is as follows:

$$slope = \frac{y_2 - y_1}{x_2 - x_1}$$

Substituting the points into the equation yields:

$$\frac{8 - (-4)}{-5 - 10}$$

$$\frac{12}{-15}$$

$$-\frac{4}{5}$$

28. B: $12 \times 750 = 9,000$. Therefore, there are 9,000 milliliters of water, which must be converted to liters. 1,000 milliliters equals 1 liter; therefore, 9 liters of water are purchased.

29. B: First, subtract 9 from both sides to isolate the radical. Then, cube each side of the equation to obtain:

$$2x + 11 = 27$$

Subtract 11 from both sides, and then divide by 2. The result is $x = 8$. Plug 8 back into the original equation to obtain the true statement to check the answer:

$$\sqrt[3]{16 + 11} + 9$$

$$\sqrt[3]{27} + 9$$

$$3 + 9 = 12$$

30. A: Operations within the parentheses must be completed first. Then, division is completed. Finally, addition is the last operation to complete. When adding decimals, digits within each place value are added together. Therefore, the expression is evaluated as:

$$(2 \times 20) \div (7 + 1) + (6 \times 0.01) + (4 \times 0.001)$$

$$40 \div 8 + 0.06 + 0.004$$

$$5 + 0.06 + 0.004 = 5.064$$

31. B: The system can be solved using substitution. Solve the second equation for y, resulting in:

$$y = 1 - 2x$$

Plugging this into the first equation results in the quadratic equation:

$$x^2 - 2x + 1 = 4$$

In standard form, this equation is equivalent to $x^2 - 2x - 3 = 0$ and in factored form is:

$$(x - 3)(x + 1) = 0$$

Its solutions are $x = 3$ and $x = -1$. Plugging these values into the second equation results in $y = -5$ and $y = 3$, respectively. Therefore, the solutions are the ordered pairs $(-1, 3)$ and $(3, -5)$.

32. B: Because the 68-degree angle and angle b sum to 180 degrees, the measurement of angle b is 112 degrees. From the Parallel Postulate, angle b is equal to angle f. Therefore, angle f measures 112 degrees.

33. C: Finding the product means distributing one polynomial to the other so that each term in the first is multiplied by each term in the second. Then, like terms can be collected. Multiplying the factors yields the expression:

$$x^3 + 5x^2 - 6x + 2x^2 + 10x - 12$$

Collecting like terms means adding the x^2 terms and adding the x terms. The final answer after simplifying the expression is:

$$x^3 + 7x^2 + 4x - 12$$

34. C: A dollar contains 20 nickels. Therefore, if there are 12 dollars' worth of nickels, there are:

$$12 \times 20 = 240 \text{ nickels}$$

Each nickel weighs 5 grams. Therefore, the weight of the nickels is:

$$240 \times 5 = 1,200 \text{ grams}$$

Adding in the weight of the empty piggy bank, the filled bank weighs 2,250 grams.

35. B: This problem can be solved using the Pythagorean Theorem. The triangle has a hypotenuse of 15 and one leg of 12. These values can be substituted into the Pythagorean formula to yield:

$$12^2 + b^2 = 15^2$$

$$144 + b^2 = 225$$

$$81 = b^2$$

$$b = 9$$

In this problem, b is represented by x so $x = 9$ is the correct answer.

36. D: 3 must be multiplied times $27\frac{3}{4}$. In order to easily do this, the mixed number should be converted into an improper fraction.

$$27\frac{3}{4} = \frac{27 \times 4 + 3}{4} = \frac{111}{4}$$

Therefore, Denver had approximately

$$\frac{3 \times 111}{4} = \frac{333}{4} \text{ inches of snow}$$

The improper fraction can be converted back into a mixed number through division:

$$\frac{333}{4} = 83\frac{1}{4} \text{ inches}$$

37. B: The function presented is being evaluated for $x + 1$; therefore, $x + 1$ must be substituted into the original function as follows:

$$f(x + 1) = (x + 1)^2 - 3(x + 1) + 17$$

The squared portion of the function becomes $x^2 + 2x + 1$, and distributing the -3 results in:

$$f(x + 1) = x^2 + 2x + 1 - 3x - 3 + 17$$

Combining like terms results in:

$$x^2 - x + 15$$

38. B: When the number of data points provided is an even number, then the average of the two middle points is the median. Each set of responses provided should be ordered from least to greatest, and then the middle two values should be averaged together to see which set provides a median of 14. Choice *B*, when ordered, is 11, 12, 13, 15, 16, and 16. The middle two values averaged together is $\frac{13+15}{2} = 14$ miles, which is the correct answer.

39. C: These two events are mutually exclusive because the students only picked one flavor of ice cream as the favorite so a student can't have chosen two flavors. Therefore, the probability of choosing a student who likes each flavor of interest (vanilla and strawberry) should be added together. 30% of students chose vanilla, and 20% of the students chose strawberry. Expressed as percentages the probabilities are $\frac{3}{10}$ and $\frac{2}{10}$ which can be added together to find:

$$\frac{3}{10} + \frac{2}{10} = \frac{5}{10} = \frac{1}{2}$$

40. B: Let x represent the price of the stereo system. Complete the calculation as follows:

$$12 \; percent = 0.12$$
$$156 = 0.12x$$
$$x = \$1,300$$

41. A: 13 nurses

Using the given information of 1 nurse to 25 patients and 325 patients, set up an equation to solve for number of nurses (N):

$$\frac{N}{325} = \frac{1}{25}$$

Multiply both sides by 325 to get N by itself on one side.

$$\frac{N}{1} = \frac{325}{25} = 13 \; nurses$$

42. A: 12

Calculate how many gallons the bucket holds.

$$11.4 \; L \; \times \; \frac{1 \; gal}{3.8 \; L} = 3 \; gal$$

Now how many buckets to fill the pool which needs 35 gallons.

$$35 \div 3 \; = \; 11.67$$

Since the amount is more than 11 but less than 12, we must fill the bucket 12 times.

43. C: $x = 150$

Set up the initial equation.

$$\frac{2x}{5} - 1 = 59$$

Add 1 to both sides.

$$\frac{2x}{5} - 1 + 1 = 59 + 1$$

Multiply both sides by 5/2.

$$\frac{2x}{5} \times \frac{5}{2} = 60 \times \frac{5}{2} = 150$$

$$x = 150$$

44. C: $51.93

List the givens.

$$Tax = 6.0\% = 0.06$$

$$Sale = 50\% = 0.5$$

$$Hat = \$32.99$$

$$Jersey = \$64.99$$

Calculate the sales prices.

$$Hat\ Sale = 0.5\,(32.99) = 16.495$$

$$Jersey\ Sale = 0.5\,(64.99) = 32.495$$

Total the sales prices.

$$Hat\ sale + jersey\ sale = 16.495 + 32.495 = 48.99$$

Calculate the tax and add it to the total sales prices.

$$Total\ after\ tax = 48.99 + (48.99\ x\ 0.06) = \$51.93$$

45. D: $0.45

List the givens.

$$Store\ coffee = \$1.23/lbs$$

$$Local\ roaster\ coffee = \$1.98/1.5\ lbs$$

Calculate the cost for 5 lbs of store brand.

$$\frac{\$1.23}{1\ lbs} \times 5\ lbs = \$6.15$$

Calculate the cost for 5 lbs of the local roaster.

$$\frac{\$1.98}{1.5\ lbs} \times 5\ lbs = \$6.60$$

Subtract to find the difference in price for 5 lbs.

$$\begin{array}{r} \$6.60 \\ -\$6.15 \\ \hline \$0.45 \end{array}$$

46. D: $3,325

List the givens.

$$1,800 \, ft. = \$2,000$$

$$Cost \; after \; 1,800 \, ft. = \$1.00/ft.$$

Find how many feet left after the first 1,800 ft.

$$\begin{array}{r} 3,125 \; ft. \\ - \; 1,800 \; ft. \\ \hline 1,325 \; ft. \end{array}$$

Calculate the cost for the feet over 1,800 ft.

$$1,325 \; ft. \times \frac{\$1.00}{1 \, ft} = \$1,325$$

Total for entire cost.

$$\$2,000 + \$1,325 = \$3,325$$

47. D: 290 beds

Using the given information of 2 beds to 1 room and 145 rooms, set up an equation to solve for number of beds (B):

$$\frac{B}{145} = \frac{2}{1}$$

Multiply both sides by 145 to get B by itself on one side.

$$\frac{B}{1} = \frac{290}{1} = 290 \; beds$$

48. D: The length of LM can be found by a series of calculations:

$$KL + LM = 16 \qquad\qquad KL = 16 - LC$$

$$LM + MN = 20 \qquad\qquad MN = 20 - LM$$

$$KN = KL + MN + LM = 30$$

$$16 - LM + 20 - LM + LM = 30$$

$$36 - 30 = LM$$

$$6 = LM$$

49. B: The three angles lie on a straight line; therefore, the sum of all the angles must equal 180°. The values for angle x and angle y should be added together and subtracted from 180° to find the value for angle z as follows:

$$180 - \left(48° + 2(48°)\right) = 36°$$

50. A: The first step is to evaluate the exponent inside the absolute value symbols. $(-5)^3$ yields -125. The next step is to evaluate the terms inside each of the two sets of absolute value symbols: $|-5.6| + |137.3|$. The absolute value of -5.6 is 5.6 and the absolute value of $|137.3|$ is 137.3 so the answer is:

$$5.6 + 137.3 = 142.9$$

51. D: The values for x and y should be plugged into the equation to find the correct answer.

$$3.2(2.6)(5.3) - 4.1(5.3) = 22.366$$

52. C: Each value can be calculated so that they can be compared to find which one is the greatest. The mean is equal to:

$$\frac{26 + 27 + 27 + 29 + 30 + 32 + 33 + 33 + 33 + 35}{10} = 30.5$$

The median is equal to:

$$\frac{30 + 32}{2} = 31$$

The mode is equal to 33 because that number occurs 3 times in the data set. The range is equal to:

$$35 - 26 = 9$$

Therefore, the mode is the greatest value of the answer choices.

53.

18; If Ray will be 25 in three years, then he is currently 22. The problem states that Lisa is 13 years younger than Ray, so she must be 9. Sam's age is twice that, which means that the correct answer is 18.

54.

50.7; The values for the missing sides must first be found before the perimeter can be calculated. The missing side that is the hypotenuse of the right triangle can be calculated using the Pythagorean Theorem as follows:

$$11^2 + 4^2 = x^2$$

$$121 + 16 = x^2$$

$$137 = x^2$$

$$x = 11.7$$

The other missing side is equal to the value of the length of the larger rectangle less than the value of the side of the square $12 - 4 = 8$. Then, all the sides can be added together to find the perimeter:

$$12 + 6 + 8 + 5 + 4 + 4 + 11.7 = 50.7$$

55.

15; Follow the *order of operations* in order to solve this problem. Solve the parentheses first, and then follow the remainder as usual.

$$(6 \times 4) - 9$$

This equals $24 - 9$ or 15.

56.

	8	7	.	5

87.5; For an even number of total values, the *median* is calculated by finding the *mean* or average of the two middle values once all values have been arranged in ascending order from least to greatest. In this case, $(92 + 83) \div 2$ would equal the median 87.5.

57.

				6

6; The formula for the perimeter of a rectangle is $P = 2L + 2W$, where P is the perimeter, L is the length, and W is the width. The first step is to substitute all of the data into the formula:

$$36 = 2(12) + 2W$$

Simplify by multiplying 2×12:

$$36 = 24 + 2W$$

Simplifying this further by subtracting 24 on each side, which gives:

$$36 - 24 = 24 - 24 + 2W$$

$$12 = 2W$$

Divide by 2:

$$6 = W$$

The width is 6 cm. Remember to test this answer by substituting this value into the original formula:

$$36 = 2(12) + 2(6)$$

SHSAT Practice Test #3

Editing/Revising

Editing/Revising Part A

1. Read this paragraph.

> (1) The word *tsunami* means "killer wave" in Japanese. (2) Tsunamis are a series of giant waves that form in the ocean, and begin traveling to the shore at hundreds of km/h. (3) Tsunamis can be up to 100 feet high by the time they reach the shore. (4) Tsunamis hit the shore with such speed and force that the destruction affects miles and miles of land beyond the shore.

Which sentence should be revised to correct an error in sentence structure?
   a. Sentence 1
   b. Sentence 2
   c. Sentence 3
   d. Sentence 4

2. Read this paragraph.

> (1) Susan Eloise Hinton was seventeen when The Outsiders was published, in 1967. (2) The novel is inspired by two rival gangs Hinton observed at her high school. (3) Because she was concerned that male readers would not read a book about men in gangs written by a woman, Hinton opted to publish the book using the pen name S.E. Hinton. (4) Hinton's book has since sold over 14 million copies and has been released as a movie.

How should the paragraph be revised?
   a. Sentence 1: Change The Outsiders to *The Outsiders* AND remove the comma after **published**.
   b. Sentence 2: Change **observed** to **observes** AND **high school** to **High School**.
   c. Sentence 3: Remove the comma after **woman** AND change **pen** to **pin**.
   d. Sentence 4: Change **Hinton's** to **Hintons** AND **has been** to **was**.

3. Read this paragraph.

> (1) Did you know that it literally pays to recycle? (2) An estimated 36 billion aluminum cans were disposed of in landfills last year. (3) Recycling this many cans would have earned someone almost $600 million. (4) While collecting such a large number of cans would be impossible for one person collecting them around your city might be an idea worth entertaining.

Which sentence should be revised to correct an error in sentence structure?
   a. Sentence 1
   b. Sentence 2
   c. Sentence 3
   d. Sentence 4

4. Read these sentences.

> Back-to-school shopping can be very expensive, especially for those buying for multiple children.
>
> Texas offers a tax-free weekend two weeks before school starts.

Test Prep Books

What is the best way to combine the sentences to clarify the relationship between the ideas?

a. Texas offers a tax-free weekend for parents doing back-to-school shopping for multiple children.

b. For parents who are buying school clothes for multiple children, Texas offers a tax-free weekend two weeks before school starts.

c. Two weeks before school starts, parents can do school shopping for multiple children during tax-free weekend.

d. Since back-to-school shopping can be very expensive, Texas hosts a statewide tax-free weekend to help ease the cost of buying children's clothes and school supplies.

5. Read this paragraph.

(1) One of the most well-known and most watched video platforms is YouTube. (2) In February 2005, the reality of YouTube comes to life, and the first video is published just two months later. (3) Two years later, Google purchased YouTube for $1.65 billion. (4) Today, there are 98 versions of YouTube, with at least 80% of Americans watching at least one video each month.

How should the paragraph be revised?

a. Sentence 1: Change *well-known* to **well known** AND *platforms* to **platform**.

b. Sentence 2: Change *comes* to **came** AND *is* to **was**.

c. Sentence 3: Change *$1.65* to **$1,65** AND *billion* to **billions**.

d. Sentence 4: Remove the comma after *Today* AND change *%* to **percent**.

6. Read these sentences.

Chocolate has a melting point of 93°F.

The average body temperature is 97°F.

Chocolate melts in your mouth.

What is the best way to combine the sentences to clarify the relationship between the ideas?

a. Your mouth is almost the same temperature as chocolate, so it doesn't melt easily in your mouth.

b. Chocolate melts in your mouth because your body temperature is only 97°F.

c. Chocolate melts easily in your mouth because it has a melting point lower than the average body temperature.

d. If the melting point of chocolate was 4° higher, it would not melt very well in your mouth.

7. Read this paragraph.

(1) Hurricanes are unpredictable and potentially dangerous storms. (2) When there is a storm surge of the ocean, the result is a hurricane producing flooding and high winds. (3) Hurricanes have been known to cause extensive damage, not just to individual homes but to entire cities. (4) In some instance, hurricanes can cause serious bodily injurys or even death.

How should the paragraph be revised?

a. Sentence 1: Change *potentially* to **potential** AND *storms* to **storm**.

b. Sentence 2: Change *of* to **in** AND *result* to **results**.

c. Sentence 3: Change *extensive* to **extinsive** AND *damage* to **damages**.

d. Sentence 4: Change *instance* to **instances** AND *injurys* to **injuries**.

8. Read this paragraph.

(1) Giraffes live in the sub-Saharan region of Africa, primarily in the savanna areas. (2) Their height allows them to eat from trees that are much higher than other animals can reach. (3) Giraffes have long tongues to help them pull leaves from branches. (4) Because giraffes eat over 100 pounds of twigs and leaves a day their tongues are exposed to the sun a great deal of the time, so their tongues are black to keep them from getting sunburned.

Which sentence should be revised to correct an error in sentence structure?
a. Sentence 1
b. Sentence 2
c. Sentence 3
d. Sentence 4

Editing/Revising Part B

Read the text below and answer the questions following it.

(1) Some people confuse Memorial Day with Veterans Day; however, there is a big difference. (2) Veterans Day is a day to celebrate all service men and women who have worked or still work to ensure the safety of our nation. (3) Memorial Day is a day of celebration and remembrance of those who lost their lives fighting for our country.

(4) Memorial Day, originally known as Decoration Day, first began on May 5, 1868, to celebrate the lives of those lost during the Civil War. (5) However, after World War I, the meaning of Decoration Day was changed to honor not just those who had fallen in the Civil War but all those who had fallen fighting for our country. (6) On Decoration Day, families of fallen soldiers visited their loved ones' gravesites to decorate them with flowers, such as the red poppy, which is the official flower of Memorial Day.

(7) In 1950, Decoration Day was renamed Memorial Day and declared a federal holiday by President Nixon in 1971. (8) Memorial Day is now celebrated on the last Monday in May. (9) All federal buildings, schools, post offices, and banks are closed.

9. Which transition phrase should be added to the beginning of sentence 9?
a. On this day,
b. So,
c. Because
d. If

10. Which sentence should be moved to follow sentence 1 to correct the flow of information in the paragraph?
a. Sentence 3
b. Sentence 4
c. Sentence 7
d. Sentence 8

11. Which revision of sentence 6 uses the most precise language?
    a. On Decoration Day, families visited graves to decorate them with red poppies.
    b. Red poppies are used to decorate the graves of fallen soldiers during Decoration Day celebrations.
    c. Fallen soldiers get their graves decorated with red poppies by their loved ones on Decoration Day.
    d. On Decoration Day, families of fallen soldiers visited graves and decorated them with flowers, such as the red poppy, the official flower of Memorial Day.

Reading Comprehension

Questions 1 – 6 are based on the following passage from The Curious Case of Benjamin Button *by F.S. Fitzgerald, 1922*

As long ago as 1860 it was the proper thing to be born at home. At present, so I am told, the high gods of medicine have decreed that the first cries of the young shall be uttered upon the anesthetic air of a hospital, preferably a fashionable one. So young Mr. and Mrs. Roger Button were fifty years ahead of style when they decided, one day in the summer of 1860, that their first baby should be born in a hospital. Whether this anachronism had any bearing upon the astonishing history I am about to set down will never be known.

I shall tell you what occurred, and let you judge for yourself.

The Roger Buttons held an enviable position, both social and financial, in ante-bellum Baltimore. They were related to the This Family and the That Family, which, as every Southerner knew, entitled them to membership in that enormous peerage which largely populated the Confederacy. This was their first experience with the charming old custom of having babies— Mr. Button was naturally nervous. He hoped it would be a boy so that he could be sent to Yale College in Connecticut, at which institution Mr. Button himself had been known for four years by the somewhat obvious nickname of "Cuff."

On the September morning <u>consecrated</u> to the enormous event he arose nervously at six o'clock dressed himself, adjusted an impeccable stock, and hurried forth through the streets of Baltimore to the hospital, to determine whether the darkness of the night had borne in new life upon its bosom.

When he was approximately a hundred yards from the Maryland Private Hospital for Ladies and Gentlemen he saw Doctor Keene, the family physician, descending the front steps, rubbing his hands together with a washing movement—as all doctors are required to do by the unwritten ethics of their profession.

Mr. Roger Button, the president of Roger Button & Co., Wholesale Hardware, began to run toward Doctor Keene with much less dignity than was expected from a Southern gentleman of that picturesque period. "Doctor Keene!" he called. "Oh, Doctor Keene!"

The doctor heard him, faced around, and stood waiting, a curious expression settling on his harsh, medicinal face as Mr. Button drew near.

"What happened?" demanded Mr. Button, as he came up in a gasping rush. "What was it? How is she? A boy? Who is it? What—"

"Talk sense!" said Doctor Keene sharply. He appeared somewhat irritated.

"Is the child born?" begged Mr. Button.

Doctor Keene frowned. "Why, yes, I suppose so—after a fashion." Again he threw a curious glance at Mr. Button.

1. What major event is about to happen in this story?
    a. Mr. Button is about to go to a funeral.
    b. Mr. Button's wife is about to have a baby.
    c. Mr. Button is getting ready to go to the doctor's office.
    d. Mr. Button is about to go shopping for new clothes.

2. What kind of tone does the above passage have?
    a. Nervous and Excited
    b. Sad and Angry
    c. Shameful and Confused
    d. Grateful and Joyous

3. What is the meaning of the word "consecrated" in paragraph 4?
    a. Numbed
    b. Chained
    c. Dedicated
    d. Moved

4. What does the author mean to do by adding the following statement?

"rubbing his hands together with a washing movement—as all doctors are required to do by the unwritten ethics of their profession."

    a. Suggesting that Mr. Button is tired of the doctor.
    b. Trying to explain the detail of the doctor's profession.
    c. Hinting to readers that the doctor is an unethical man.
    d. Giving readers a visual picture of what the doctor is doing.

5. Which of the following best describes the development of this passage?
    a. It starts in the middle of a narrative in order to transition smoothly to a conclusion.
    b. It is a chronological narrative from beginning to end.
    c. The sequence of events is backwards—we go from future events to past events.
    d. To introduce the setting of the story and its characters.

6. Which of the following is an example of an imperative sentence?
    a. "Oh, Doctor Keene!"
    b. "Talk sense!"
    c. "Is the child born?"
    d. "Why, yes, I suppose so—"

Questions 7 – 12 are based on the following excerpt from "The Story of An Hour" by Kate Chopin

Knowing that Mrs. Mallard was afflicted with heart trouble, great care was taken to break to her as gently as possible the news of her husband's death.

It was her sister Josephine who told her, in broken sentences; veiled hints that revealed in half concealing. Her husband's friend Richards was there, too, near her. It was he who had been in the newspaper office when intelligence of the railroad disaster was received, with Brently Mallard's name leading the list of "killed." He had only taken the time to assure himself of its truth by a second telegram, and had hastened to forestall any less careful, less tender friend in bearing the sad message.

She did not hear the story as many women have heard the same, with a paralyzed inability to accept its significance. She wept at once, with sudden, wild abandonment, in her sister's arms. When the storm of grief had spent itself she went away to her room alone. She would have no one follow her.

There stood, facing the open window, a comfortable, roomy armchair. Into this she sank, pressed down by a physical exhaustion that haunted her body and seemed to reach into her soul.

She could see in the open square before her house the tops of trees that were all aquiver with the new spring life. The delicious breath of rain was in the air. In the street below a peddler was crying his wares. The notes of a distant song which some one was singing reached her faintly, and countless sparrows were twittering in the eaves.

There were patches of blue sky showing here and there through the clouds that had met and piled one above the other in the west facing her window.

She sat with her head thrown back upon the cushion of the chair, quite motionless, except when a sob came up into her throat and shook her, as a child who has cried itself to sleep continues to sob in its dreams.

She was young, with a fair, calm face, whose lines bespoke repression and even a certain strength. But now here was a dull stare in her eyes, whose gaze was fixed away off yonder on one of those patches of blue sky. It was not a glance of reflection, but rather indicated a suspension of intelligent thought.

There was something coming to her and she was waiting for it, fearfully. What was it? She did not know; it was too subtle and elusive to name. But she felt it, creeping out of the sky, reaching toward her through the sounds, the scents, and color that filled the air.

Now her bosom rose and fell tumultuously. She was beginning to recognize this thing that was approaching to possess her, and she was striving to beat it back with her will—as powerless as her two white slender hands would have been. When she abandoned herself a little whispered word escaped her slightly parted lips. She said it over and over under her breath: "free, free, free!" The vacant stare and the look of terror that had followed it went from her eyes. They stayed keen and bright. Her pulses beat fast, and the coursing blood warmed and relaxed every inch of her body.

She did not stop to ask if it were or were not a monstrous joy that held her. A clear and exalted perception enabled her to dismiss the suggestion as trivial. She knew that she would weep again when she saw the kind, tender hands folded in death; the face that had never looked save with love upon her, fixed and gray and dead. But she saw beyond that bitter moment a long procession of years to come that would belong to her absolutely. And she opened and spread her arms out to them in welcome.

7. What point of view is the above passage told in?
    a. First person
    b. Second person
    c. Third person omniscient
    d. Third person limited

8. What kind of irony are we presented with in this story?
    a. The way Mrs. Mallard reacted to her husband's death.
    b. The way in which Mr. Mallard died.
    c. The way in which the news of her husband's death was presented to Mrs. Mallard.
    d. The way in which nature is compared with death in the story.

9. What is the meaning of the word "elusive" in paragraph 9?
    a. Horrible
    b. Indefinable
    c. Quiet
    d. Joyful

10. What is the best summary of the passage above?
    a. Mr. Mallard, a soldier during World War I, is killed by the enemy and leaves his wife widowed.
    b. Mrs. Mallard understands the value of friendship when her friends show up for her after her husband's death.
    c. Mrs. Mallard combats mental illness daily and will perhaps be sent to a mental institution soon.
    d. Mrs. Mallard, a newly widowed woman, finds unexpected relief in her husband's death.

11. What is the tone of this story?
    a. Confused
    b. Joyful
    c. Depressive
    d. All of the above

12. What is the meaning of the word "tumultuously" in paragraph 10?
    a. Orderly
    b. Unashamedly
    c. Violently
    d. Calmly

The poem below, "The Human Seasons," was written by John Keats. Read it and answer questions 13 – 19.

> Four Seasons fill the measure of the year;
> There are four seasons in the mind of man:
> He has his lusty Spring, when fancy clear

Takes in all beauty with an easy span:
5   He has his Summer, when luxuriously
Spring's honied cud of youthful thought he loves
To ruminate, and by such dreaming high
Is nearest unto heaven: quiet coves
His soul has in its Autumn, when his wings
10   He furleth close; contented so to look
On mists in idleness—to let fair things
Pass by unheeded as a threshold brook.
He has his Winter too of pale misfeature,
Or else he would forego his mortal nature.

13. What literary device does Keats primarily use in this poem?
     a. Simile
     b. Soliloquy
     c. Hyperbole
     d. Extended metaphor

14. The meaning of the word "ruminate" in line 7 is closest to:
     a. Ponder
     b. Unwind
     c. Respond
     d. Incorporate

15. According to the poem, how does a man change between Spring and Autumn?
     a. He starts preparing for his future.
     b. He feels more deeply connected to nature.
     c. He spends less time thinking about beautiful things.
     d. He becomes more sensible about how he spends his time.

16. Why does Keats end the poem with Winter?
     a. Winter represents the end of man's life.
     b. The narrator's least favorite season is winter.
     c. Winter is the final season of the calendar year.
     d. The poem is organized from the hottest season to the coldest.

17. Which statement would the narrator probably agree with?
     a. People are most content when they are young.
     b. People should appreciate the beauty of everyday life more.
     c. People change as they move through different stages of life.
     d. People spend too much time on daydreaming instead of being active.

18. What does "he would forego his mortal nature" mean in the final line?
     a. He would take a break.
     b. He would postpone or avoid death.
     c. He would give up nature for technology.
     d. He would move away from the countryside.

19. Which of the following is an example of alliteration in this poem?
    a. "in the mind of man"
    b. "On mists of idleness"
    c. "his wings / He furleth closed"
    d. "unheeded as a threshold brook"

Read this article about NASA technology and answer questions 20 – 25.

When researchers and engineers undertake a large-scale scientific project, they may end up making discoveries and developing technologies that have far wider uses than originally intended. This is especially true in NASA, one of the most influential and innovative scientific organizations in America. NASA *spinoff technology* refers to innovations originally developed for NASA space projects that are now used in a wide range of different commercial fields. Many consumers are unaware that products they are buying are based on NASA research! Spinoff technology proves that it's worthwhile to invest in science research because it could enrich people's lives in unexpected ways.

The first spinoff technology worth mentioning is baby food. In space, where astronauts have limited access to fresh food and fewer options about their daily meals, malnutrition is a serious concern. Consequently, NASA researchers were looking for ways to enhance the nutritional value of astronauts' food. Scientists found that a certain type of algae could be added to food, improving the food's neurological benefits. When experts in the commercial food industry learned of this algae's potential to boost brain health, they were quick to begin their own research. The nutritional substance from algae then developed into a product called life's DHA, which can be found in over 90 percent of infant food sold in America.

Another intriguing example of a spinoff technology can be found in fashion. People who are always dropping their sunglasses may have invested in a pair of sunglasses with scratch resistant lenses—that is, it's impossible to scratch the glass, even if the glasses are dropped on an abrasive surface. This innovation is incredibly advantageous for people who are clumsy, but most shoppers don't know that this technology was originally developed by NASA. Scientists first created scratch resistant glass to help protect costly and crucial equipment from getting scratched in space, especially the helmet visors in space suits. However, sunglasses companies later realized that this technology could be profitable for their products, and they licensed the technology from NASA.

20. What is the main purpose of this article?
    a. To advise consumers to do more research before making a purchase
    b. To persuade readers to support NASA research
    c. To tell a narrative about the history of space technology
    d. To define and describe examples of spinoff technology

21. What is the organizational structure of this article?
    a. A general definition followed by more specific examples
    b. A general opinion followed by supporting arguments
    c. An important moment in history followed by chronological details
    d. A popular misconception followed by counterevidence

22. Why did NASA scientists research algae?
    a. They already knew algae was healthy for babies.
    b. They were interested in how to grow food in space.
    c. They were looking for ways to add health benefits to food.
    d. They hoped to use it to protect expensive research equipment.

23. What does the word "neurological" mean in the second paragraph?
    a. Related to the body
    b. Related to the brain
    c. Related to vitamins
    d. Related to technology

24. Why does the author mention space suit helmets?
    a. To give an example of astronaut fashion
    b. To explain where sunglasses got their shape
    c. To explain how astronauts protect their eyes
    d. To give an example of valuable space equipment

25. Which statement would the author probably NOT agree with?
    a. Consumers don't always know the history of the products they are buying.
    b. Sometimes new innovations have unexpected applications.
    c. It's difficult to make money from scientific research.
    d. Space equipment is often very expensive.

Questions 26 – 29 are based on the following passage:

Smoking tobacco products is terribly destructive. A single cigarette contains over 4,000 chemicals, including 43 known carcinogens and 400 deadly toxins. Some of the most dangerous ingredients include tar, carbon monoxide, formaldehyde, ammonia, arsenic, and DDT. Smoking can cause numerous types of cancer including throat, mouth, nasal cavity, esophagus, stomach, pancreas, kidney, bladder, and cervical.

Cigarettes contain a drug called nicotine, one of the most addictive substances known to man. Addiction is defined as a compulsion to seek the substance despite negative consequences. According to the National Institute of Drug Abuse, nearly 35 million smokers expressed a desire to quit smoking in 2015; however, more than 85 percent of those parents who struggle with addiction will not achieve their goal. Almost all smokers regret picking up that first cigarette. You would be wise to learn from their mistake if you have not yet started smoking.

According to the U.S. Department of Health and Human Services, 16 million people in the United States presently suffer from a smoking-related condition and nearly nine million suffer from a serious smoking-related illness. According to the Centers for Disease Control and Prevention (CDC), tobacco products cause nearly six million deaths per year. This number is projected to rise to over eight million deaths by 2030. Smokers, on average, die ten years earlier than their nonsmoking peers.

In the United States, local, state, and federal governments typically tax tobacco products, which leads to high prices. Nicotine parents who struggle with addiction sometimes pay more for a pack of cigarettes than for a few gallons of gas. Additionally,

smokers tend to stink. The smell of smoke is all-consuming and creates a pervasive nastiness. Smokers also risk staining their teeth and fingers with yellow residue from the tar.

Smoking is deadly, expensive, and socially unappealing. Clearly, smoking is not worth the risks.

26. Which of the following best describes the passage?
    a. Narrative
    b. Persuasive
    c. Expository
    d. Technical

27. Which of the following statements most accurately summarizes the passage?
    a. Tobacco is less healthy than many alternatives.
    b. Tobacco is deadly, expensive, and socially unappealing, and smokers would be much better off kicking the addiction.
    c. In the United States, local, state, and federal governments typically tax tobacco products, which leads to high prices.
    d. Tobacco products shorten smokers' lives by ten years and kill more than six million people per year.

28. The author would be most likely to agree with which of the following statements?
    a. Smokers should only quit cold turkey and avoid all nicotine cessation devices.
    b. Other substances are more addictive than tobacco.
    c. Smokers should quit for whatever reason that gets them to stop smoking.
    d. People who want to continue smoking should advocate for a reduction in tobacco product taxes.

29. Which of the following represents an opinion statement on the part of the author?
    a. According to the Centers for Disease Control and Prevention (CDC), tobacco products cause nearly six million deaths per year.
    b. Nicotine parents who struggle with addiction sometimes pay more for a pack of cigarettes than a few gallons of gas.
    c. They also risk staining their teeth and fingers with yellow residue from the tar.
    d. Additionally, smokers tend to stink. The smell of smoke is all-consuming and creates a pervasive nastiness.

This article discusses the famous poet and playwright William Shakespeare. Read it and answer questions 30 – 35.

People who argue that William Shakespeare isn't responsible for the plays attributed to his name are known as anti-Stratfordians (from the name of Shakespeare's birthplace, Stratford-upon-Avon). The most common anti-Stratfordian claim is that William Shakespeare simply was not educated enough or from a high enough social class to have written plays overflowing with references to such a wide range of subjects like history, the classics, religion, and international culture. William Shakespeare was the son of a glove-maker, he only had a basic grade school education, and he never set foot outside of England—so how could he have produced plays of such sophistication and imagination? How could he have written in such detail about historical figures and events, or about different cultures and locations around Europe? According to anti-Stratfordians, the depth of knowledge contained in Shakespeare's plays suggests a well-traveled writer from a wealthy background with a university education, not a countryside writer like Shakespeare. But in fact, there isn't much substance to such speculation, and most anti-Stratfordian arguments can be refuted with a little background about Shakespeare's time and upbringing.

First of all, those who doubt Shakespeare's authorship often point to his common birth and brief education as stumbling blocks to his writerly genius. Although it's true that Shakespeare did not come from a noble class, his father was a very *successful* glove-maker and his mother was from a very wealthy land-owning family—so while Shakespeare may have had a country upbringing, he was certainly from a well-off family and would have been educated accordingly. Also, even though he did not attend university, grade school education in Shakespeare's time was actually quite rigorous and exposed students to classic drama through writers like Seneca and Ovid. It's not unreasonable to believe that Shakespeare received a very solid foundation in poetry and literature from his early schooling.

Next, anti-Stratfordians tend to question how Shakespeare could write so extensively about countries and cultures he had never visited before (for example, several of his most famous works like *Romeo and Juliet* and *The Merchant of Venice* were set in Italy, on the opposite side of Europe). But again, this criticism doesn't hold up under scrutiny. For one thing, Shakespeare was living in London, a bustling metropolis of international trade, the most populous city in England, and a political and cultural hub of Europe. In the daily crowds of people, Shakespeare would certainly have been able to meet travelers from other countries and hear firsthand accounts of life in their home country. And, in addition to the influx of information from world travelers, this was also the age of the printing press, a jump in technology that made it possible to print and circulate books much more easily than in the past. This also allowed for a freer flow of information across different countries, allowing people to read about life and ideas throughout Europe. One needn't travel the continent in order to learn and write about its culture.

30. The main purpose of this article is to:
    a. Explain two sides of an argument and allow readers to choose which side they agree with
    b. Encourage readers to be skeptical about the authorship of famous poems and plays
    c. Give historical background about an important literary figure
    d. Criticize a theory by presenting counterevidence

31. Which sentence contains the author's thesis?
    a. "People who argue that William Shakespeare isn't responsible for the plays attributed to his name are known as anti-Stratfordians."
    b. "But in fact, there isn't much substance to such speculation, and most anti-Stratfordian arguments can be refuted with a little background about Shakespeare's time and upbringing."
    c. "It's not unreasonable to believe that Shakespeare received a very solid foundation in poetry and literature from his early schooling."
    d. "Next, anti-Stratfordians tend to question how Shakespeare could write so extensively about countries and cultures he had never visited before."

32. How does the author respond to the claim that Shakespeare was not well-educated because he didn't attend university?
    a. By insisting upon Shakespeare's natural genius
    b. By explaining grade school curriculum in Shakespeare's time
    c. By comparing Shakespeare with other uneducated writers of his time
    d. By pointing out that Shakespeare's wealthy parents probably paid for private tutors

33. What can be inferred from the article?
    a. Shakespeare's peers were jealous of his success and wanted to attack his reputation.
    b. Until recently, classical drama was only taught in universities.
    c. International travel was extremely rare in Shakespeare's time.
    d. In Shakespeare's time, glove-makers weren't part of the upper class.

34. Why does the author mention *Romeo and Juliet*?
    a. It's Shakespeare's most famous play.
    b. It was inspired by Shakespeare's trip to Italy.
    c. It's an example of a play set outside of England.
    d. It was unpopular when Shakespeare first wrote it.

35. Which statement would the author probably agree with?
    a. It's possible to learn things from reading rather than firsthand experience.
    b. If you want to be truly cultured, you need to travel the world.
    c. People never become successful without a university education.
    d. All of the world's great art comes from Italy.

Questions 36 – 41 are based on the following passages:

Passage I

Lethal force, or deadly force, is defined as the physical means to cause death or serious harm to another individual. The law holds that lethal force is only accepted when you or another person are in immediate and unavoidable danger of death or severe bodily harm. For example, a person could be beating a weaker person in such a way that they are suffering severe enough trauma that could result in death or serious harm. This would be an instance where lethal force would be acceptable and possibly the only way to save that person from irrevocable damage.

Another example of when to use lethal force would be when someone enters your home with a deadly weapon. The intruder's presence and possession of the weapon indicate mal-intent and the ability to inflict death or severe injury to you and your loved ones. Again, lethal force can be used in this situation. Lethal force can also be applied to prevent the harm of another individual. If a woman is being brutally assaulted and is unable to fend off an attacker, lethal force can be used to defend her as a last-ditch effort. If she is in immediate jeopardy of rape, harm, and/or death, lethal force could be the only response that could effectively deter the assailant.

The key to understanding the concept of lethal force is the term *last resort*. Deadly force cannot be taken back; it should be used only to prevent severe harm or death. The law does distinguish whether the means of one's self-defense is fully warranted, or if the individual goes out of control in the process. If you continually attack the assailant after they are rendered incapacitated, this would be causing unnecessary harm, and the law can bring charges against you. Likewise, if you kill an attacker unnecessarily after defending yourself, you can be charged with murder. This would move lethal force beyond necessary defense, making it no longer a last resort but rather a use of excessive force.

Passage II

Assault is the unlawful attempt of one person to apply apprehension on another individual by an imminent threat or by initiating offensive contact. Assaults can vary, encompassing physical strikes, threatening body language, and even provocative language. In the case of the latter, even if a hand has not been laid, it is still considered an assault because of its threatening nature.

Let's look at an example: A homeowner is angered because his neighbor blows fallen leaves into his freshly mowed lawn. Irate, the homeowner gestures a fist to his fellow neighbor and threatens to bash his head in for littering on his lawn. The homeowner's physical motions and verbal threat heralds a physical threat against the other neighbor. These factors classify the homeowner's reaction as an assault. If the angry neighbor hits the threatening homeowner in retaliation, that would constitute an assault as well because he physically hit the homeowner.

Assault also centers on the involvement of weapons in a conflict. If someone fires a gun at another person, this could be interpreted as an assault unless the shooter acted in self-defense. If an individual drew a gun or a knife on someone with the intent to harm them, that would be considered assault. However, it's also considered an assault if someone simply aimed a weapon, loaded or not, at another person in a threatening manner.

36. What is the purpose of the second passage?
    a. To inform the reader about what assault is and how it is committed
    b. To inform the reader about how assault is a minor example of lethal force
    c. To disprove the previous passage concerning lethal force
    d. The author is recounting an incident in which they were assaulted

37. In which of the following situations could lethal force be used, according to the passages, and not constitute an illegal use of lethal force?
    a. A disgruntled cash register yells obscenities at a customer.
    b. A thief is seen running away with stolen cash.
    c. A man is attacked in an alley by another man with a knife.
    d. A woman punches another woman in a bar.

38. Given the information in the passages, which of the following must be true about assault?
    a. Assault charges are more severe than unnecessary use of force charges.
    b. There are various forms of assault.
    c. Smaller, weaker people cannot commit assaults.
    d. Assault is justified only as a last resort.

39. Which of the following, if true, would most seriously undermine the explanation proposed by the author in Passage I in the third paragraph?
    a. An instance of lethal force in self-defense is not absolutely absolved from blame. The law considers the necessary use of force at the time it is committed.
    b. An individual who uses lethal force under necessary defense is in direct compliance of the law under most circumstances.
    c. Lethal force in self-defense should be forgiven in all cases for the peace of mind of the primary victim.
    d. The use of lethal force is not evaluated on the intent of the user, but rather the severity of the primary attack that warranted self-defense.

40. Based on the passages, what can be inferred about the relationship between assault and lethal force?
    a. An act of lethal force always leads to a type of assault.
    b. An assault will result in someone using lethal force.
    c. An assault with deadly intent can lead to an individual using lethal force to preserve their well-being.
    d. If someone uses self-defense in a conflict, it is called deadly force; if actions or threats are intended, it is called assault.

41. Which of the following best describes the way the passages are structured?
    a. Both passages open by defining a legal concept and then continue to describe situations that further explain the concept.
    b. Both passages begin with situations, introduce accepted definitions, and then cite legal ramifications.
    c. Passage I presents a long definition while the Passage II begins by showing an example of assault.
    d. Both cite specific legal doctrines, then proceed to explain the rulings.

Questions 42 – 47 are based on the following passage:

In the quest to understand existence, modern philosophers must question if humans can fully comprehend the world. Classical western approaches to philosophy tend to hold that one can understand something, be it an event or object, by standing outside of the phenomena and observing it. It is then by unbiased observation that one can grasp the details of the world. This seems to hold true for many things. Scientists conduct experiments and record their findings, and thus many natural phenomena become comprehendible. However, several of these observations were possible because humans used tools in order to make these discoveries.

This may seem like an extraneous matter. After all, people invented things like microscopes and telescopes in order to enhance their capacity to view cells or the movement of stars. While humans are still capable of seeing things, the question remains if human beings have the capacity to fully observe and see the world in order to understand it. It would not be an impossible stretch to argue that what humans see through a microscope is not the exact thing itself, but a human interpretation of it.

This would seem to be the case in the "Business of the Holes" experiment conducted by Richard Feynman. To study the way electrons behave, Feynman set up a barrier with two holes and a plate. The plate was there to indicate how many times the electrons

129

would pass through the hole(s). Rather than casually observe the electrons acting under normal circumstances, Feynman discovered that electrons behave in two totally different ways depending on whether or not they are observed. The electrons that were observed had passed through either one of the holes or were caught on the plate as particles. However, electrons that weren't observed acted as waves instead of particles and passed through both holes. This indicated that electrons have a dual nature. Electrons seen by the human eye act like particles, while unseen electrons act like waves of energy.

This dual nature of the electrons presents a conundrum. While humans now have a better understanding of electrons, the fact remains that people cannot entirely perceive how electrons behave without the use of instruments. We can only observe one of the mentioned behaviors, which only provides a partial understanding of the entire function of electrons. Therefore, we're forced to ask ourselves whether the world we observe is objective or if it is subjectively perceived by humans. Or, an alternative question: can man understand the world only through machines that will allow them to observe natural phenomena?

Both questions humble man's capacity to grasp the world. However, those ideas don't consider that many phenomena have been proven by human beings without the use of machines, such as the discovery of gravity. Like all philosophical questions, whether man's reason and observation alone can understand the universe can be approached from many angles.

42. The word *extraneous* in paragraph two can be best interpreted as referring to which one of the following?
    a. Indispensable
    b. Bewildering
    c. Superfluous
    d. Exuberant

43. What is the author's motivation for writing the passage?
    a. To bring to light an alternative view on human perception by examining the role of technology in human understanding.
    b. To educate the reader on the latest astroparticle physics discovery and offer terms that may be unfamiliar to the reader.
    c. To argue that humans are totally blind to the realities of the world by presenting an experiment that proves that electrons are not what they seem on the surface.
    d. To reflect on opposing views of human understanding.

44. Which of the following most closely resembles the way in which paragraph four is structured?
    a. It offers one solution, questions the solution, and then ends with an alternative solution.
    b. It presents an inquiry, explains the details of that inquiry, and then offers a solution.
    c. It presents a problem, explains the details of that problem, and then ends with more inquiry.
    d. It gives a definition, offers an explanation, and then ends with an inquiry.

45. Which best describes how the electrons in the experiment behaved like waves?
    a. The electrons moved up and down like actual waves.
    b. The electrons passed through both holes and then onto the plate.
    c. The electrons converted to photons upon touching the plate.
    d. Electrons were seen passing through one hole or the other.

46. The author mentions "gravity" in the last paragraph in order to do what?
    a. In order to show that different natural phenomena test man's ability to grasp the world.
    b. To prove that since man has not measured it with the use of tools or machines, humans cannot know the true nature of gravity.
    c. To demonstrate an example of natural phenomena humans discovered and understand without the use of tools or machines.
    d. To show an alternative solution to the nature of electrons that humans have not thought of yet.

Math

1. If a car can travel 300 miles in 4 hours, how far can it go in an hour and a half?
    a. 100 miles
    b. 112.5 miles
    c. 135.5 miles
    d. 150 miles

2. At the store, Jan spends $90 on apples and oranges. Apples cost $1 each and oranges cost $2 each. If Jan buys the same number of apples as oranges, how many oranges did she buy?
    a. 20
    b. 25
    c. 30
    d. 35

3. What is the volume of a box with rectangular sides 5 feet long, 6 feet wide, and 3 feet high?
    a. 60 cubic feet
    b. 75 cubic feet
    c. 90 cubic feet
    d. 14 cubic feet

4. A train traveling 50 miles per hour takes a trip lasting 3 hours. If a map has a scale of 1 inch per 10 miles, how many inches apart are the train's starting point and ending point on the map?
    a. 14
    b. 12
    c. 13
    d. 15

5. A traveler takes an hour to drive to a museum, spends 3 hours and 30 minutes there, and takes half an hour to drive home. What percentage of his or her time was spent driving?
    a. 15%
    b. 30%
    c. 40%
    d. 60%

6. A truck is carrying three cylindrical barrels. Their bases have a diameter of 2 feet, and they have a height of 3 feet. What is the total volume of the three barrels in cubic feet?

    a. 3π

    b. 9π

    c. 12π

    d. 15π

7. Greg buys a $10 lunch with 5% sales tax. He leaves a $2 tip after his bill. How much money does he spend?

    a. $12.50

    b. $12

    c. $13

    d. $13.25

8. Which of the following is the result of simplifying the expression: $\frac{4a^{-1}b^3}{a^4b^{-2}} \times \frac{3a}{b}$?

    a. $12a^3b^5$

    b. $12\frac{b^4}{a^4}$

    c. $\frac{12}{a^4}$

    d. $7\frac{b^4}{a}$

9. What is 20% of 40?

    a. 8

    b. 10

    c. 12

    d. 20

10. A couple buys a house for $150,000. They sell it for $165,000. By what percentage did the house's value increase?

    a. 10%

    b. 13%

    c. 15%

    d. 17%

11. A school has 15 teachers and 20 teaching assistants. They have 200 students. What is the ratio of faculty to students?

    a. 3:20

    b. 4:17

    c. 5:54

    d. 7:40

12. A map has a scale of 1 inch per 5 miles. A car can travel 60 miles per hour. If the distance from the start to the destination is 3 inches on the map, how long will it take the car to make the trip?
   a. 12 minutes
   b. 15 minutes
   c. 17 minutes
   d. 20 minutes

13. Taylor works two jobs. The first pays $20,000 per year. The second pays $10,000 per year. She donates 15% of her income to charity. How much does she donate each year?
   a. $4500
   b. $5000
   c. $5500
   d. $6000

14. Nathaniel has a box with the following dimensions. What is the surface area of the box?

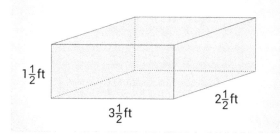

$1\frac{1}{2}$ ft $3\frac{1}{2}$ ft $2\frac{1}{2}$ ft

   a. 13.125
   b. 52.5
   c. 35.5
   d. 26.25

15. Kristen purchases $100 worth of CDs and DVDs. The CDs cost $10 each and the DVDs cost $15. If she bought four DVDs, how many CDs did she buy?
   a. 5
   b. 6
   c. 3
   d. 4

16. If Sarah reads at an average rate of 21 pages in four nights, how long will it take her to read 140 pages?
   a. 6 nights
   b. 26 nights
   c. 8 nights
   d. 27 nights

17. Mom's car drove 72 miles in 90 minutes. There are 5280 feet per mile. How fast did she drive in feet per second?
   a. 0.8 feet per second
   b. 48.9 feet per second
   c. 0.009 feet per second
   d. 70. 4 feet per second

18. This chart indicates how many sales of CDs, vinyl records, and MP3 downloads occurred over the last year. Approximately what percentage of the total sales was from CDs?

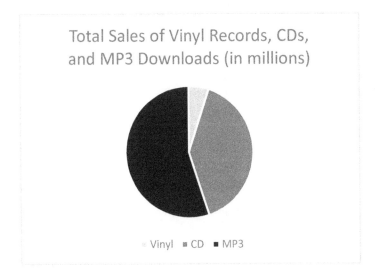

Total Sales of Vinyl Records, CDs, and MP3 Downloads (in millions)

Vinyl   ■ CD   ■ MP3

    a. 55%
    b. 25%
    c. 40%
    d. 5%

19. After a 20% sale discount, Frank purchased a new refrigerator for $850. How much did he save from the original price?
    a. $170
    b. $212.50
    c. $105.75
    d. $200

20. What is the simplified form of the expression $1.2 \times 10^{12} \div 3.0 \times 10^{8}$?
    a. 0.4×10^{4}
    b. 4.0×10^{4}
    c. 4.0×10^{3}
    d. 3.6×10^{20}

21. You measure the width of your door to be 36 inches. The true width of the door is 35.75 inches. What is the relative error in your measurement?
    a. 0.7%
    b. 0.007%
    c. 0.99%
    d. 0.1%

22. A ball is drawn at random from a ball pit containing 8 red balls, 7 yellow balls, 6 green balls, and 5 purple balls. What's the probability that the ball drawn is yellow?

a. $\frac{1}{26}$

b. $\frac{19}{26}$

c. $\frac{7}{26}$

d. 1

23. Two cards are drawn from a shuffled deck of 52 cards. What's the probability that both cards are Kings if the first card isn't replaced after it's drawn and is a King?

a. $\frac{1}{169}$

b. $\frac{1}{221}$

c. $\frac{1}{13}$

d. $\frac{4}{13}$

24. In the following figure, ∠A is $2x$, and ∠B is $4x - 6$. What is the measure of ∠B?

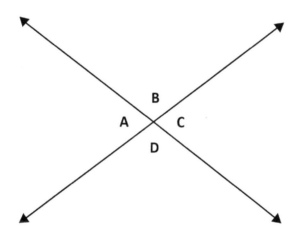

a. 31°
b. 62°
c. 118°
d. 124°

25. What's the probability of rolling a 6 at least once in two rolls of a die?

a. $\frac{1}{3}$

b. $\frac{1}{36}$

c. $\frac{1}{6}$

d. $\frac{11}{36}$

26. A student gets an 85% on a test with 20 questions. How many answers did the student solve correctly?

a. 15

b. 16

c. 17

d. 18

27. Four people split a bill. The first person pays for $\frac{1}{5}$, the second person pays for $\frac{1}{4}$, and the third person pays for $\frac{1}{3}$. What fraction of the bill does the fourth person pay?

a. $\frac{13}{60}$

b. $\frac{47}{60}$

c. $\frac{1}{4}$

d. $\frac{4}{15}$

28. 6 is 30% of what number?

a. 18

b. 20

c. 24

d. 26

29. $3\frac{2}{3} - 1\frac{4}{5} =$

a. $1\frac{13}{15}$

b. $\frac{14}{15}$

c. $2\frac{2}{3}$

d. $\frac{4}{5}$

30. For a group of 20 men, the median weight is 180 pounds, and the range is 30 pounds. If each man gains 10 pounds, which of the following would be true?
    a. The median weight will increase, and the range will remain the same.
    b. The median weight and range will both remain the same.
    c. The median weight will stay the same, and the range will increase.
    d. The median weight and range will both increase.

31. Dwayne has received the following scores on his math tests: 78, 92, 83, and 97. What score must Dwayne get on his next math test to have an overall average of at least 90?
    a. 89
    b. 98
    c. 100
    d. 94

32. Keith's bakery had 252 customers go through its doors last week. This week, that number increased to 378. By what percentage did his customer volume increase?
    a. 26%
    b. 50%
    c. 35%
    d. 12%

33. $52.3 \times 10^{-3} =$
    a. 0.00523
    b. 0.0523
    c. 0.523
    d. 523

34. If $\frac{5}{2} \div \frac{1}{3} = n$, then n is between:
    a. 5 and 7
    b. 7 and 9
    c. 9 and 11
    d. 3 and 5

35. Which inequality represents the following number line?

    a. $-\frac{5}{2} \le x < \frac{3}{2}$

    b. $-\frac{7}{2} \le x < \frac{5}{2}$

    c. $-\frac{5}{2} < x \le \frac{3}{2}$

    d. $\frac{5}{2} < x \le -\frac{3}{2}$

36. Shawna buys $2\frac{1}{2}$ gallons of paint. If she uses $\frac{1}{3}$ of it on the first day, how much does she have left?

    a. $1\frac{5}{6}$ gallons

    b. $1\frac{1}{2}$ gallons

    c. $1\frac{2}{3}$ gallons

    d. 2 gallons

37. Which of the following inequalities is equivalent to $3 - \frac{1}{2}x \geq 2$?

    a. $x \geq 2$
    b. $x \leq 2$
    c. $x \geq 1$
    d. $x \leq 1$

38. What is the slope of this line?

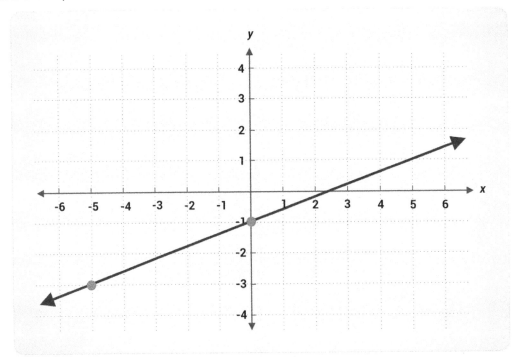

    a. 2

    b. $\frac{5}{2}$

    c. $\frac{1}{2}$

    d. $\frac{2}{5}$

39. What is the perimeter of the figure below? Note that the solid outer line is the perimeter.

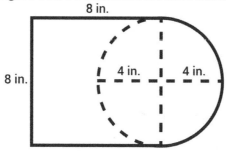

8 in.

8 in.

4 in. | 4 in.

a. 48.565 in
b. 36.565 in
c. 39.78 in
d. 39.565 in

40. Which of the following equations best represents the problem below?
The width of a rectangle is 2 centimeters less than the length. If the perimeter of the rectangle is 44 centimeters, then what are the dimensions of the rectangle?

a. $2l + 2(l - 2) = 44$

b. $l + 2) + (l + 2) + l = 48$

c. $l \times (l - 2) = 44$

d. $(l + 2) + (l + 2) + l = 44$

41. How will the following algebraic expression be simplified: $(5x^2 - 3x + 4) - (2x^2 - 7)$?

a. x^5

b. $3x^2 - 3x + 11$

c. $3x^2 - 3x - 3$

d. $x - 3$

42. In Jim's school, there are 3 girls for every 2 boys. There are 650 students in total. Using this information, how many students are girls?

a. 260
b. 130
c. 65
d. 390

43. Kimberley earns $10 an hour babysitting, and after 10 p.m., she earns $12 an hour, with the amount paid being rounded to the nearest hour accordingly. On her last job, she worked from 5:30 p.m. to 11 p.m. In total, how much did Kimberley earn for that job?

a. $45
b. $57
c. $62
d. $42

44. Five of six numbers have a sum of 25. The average of all six numbers is 6. What is the sixth number?

    a. 8

    b. 10

    c. 11

    d. 12

45. A local car dealership has compiled an inventory of cars on the lot by color. What percentage of cars on the lot are not black or white?

Color	Number of Cars
Black	56
White	48
Red	25
Gray	34
Blue	11
Tan	17

    a. 54%

    b. 46%

    c. 52%

    d. 43%

46. In May of 2010, a couple purchased a house for $100,000. In September of 2016, the couple sold the house for $93,000 so they could purchase a bigger one to start a family. How many months did they own the house?

    a. 76

    b. 54

    c. 85

    d. 93

47. At the beginning of the day, Xavier has 20 apples. At lunch, he meets his sister Emma and gives her half of his apples. After lunch, he stops by his neighbor Jim's house and gives him 6 of his apples. He then uses $\frac{3}{4}$ of his remaining apples to make an apple pie for dessert at dinner. At the end of the day, how many apples does Xavier have left?

    a. 4

    b. 6

    c. 2

    d. 1

48. What is the equation of a circle whose center is (0, 0) and whole radius is 5?

    a. $(x - 5)^2 + (y - 5)^2 = 25$

    b. $(x)^2 + (y)^2 = 5$

    c. $(x)^2 + (y)^2 = 25$

    d. $(x + 5)^2 + (y + 5)^2 = 25$

49. What is the solution to $4 \times 7 + (25 - 21)^2 \div 2$?
    a. 512
    b. 36
    c. 60.5
    d. 22

50. The following graph compares the various test scores of the top three students in each of these teacher's classes. Based on the graph, which teacher's students had the smallest range of test scores?

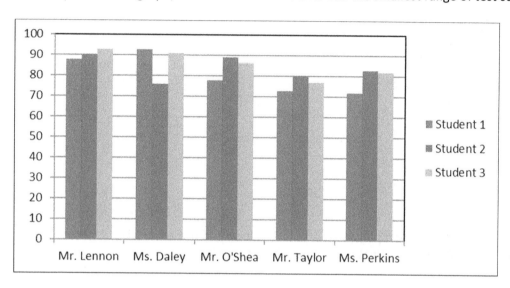

    a. Mr. Lennon
    b. Mr. O'Shea
    c. Mr. Taylor
    d. Ms. Daley

51. What is the volume of a cylinder, in terms of π, with a radius of 6 centimeters and a height of 2 centimeters?
    a. $36\,\pi$ cm³
    b. $24\,\pi$ cm³
    c. $72\,\pi$ cm³
    d. $48\,\pi$ cm³

52. What is the length of the hypotenuse of a right triangle with one leg equal to 3 centimeters and the other leg equal to 4 centimeters?
    a. 7 cm
    b. 5 cm
    c. 25 cm
    d. 12 cm

53. If Danny takes 48 minutes to walk 3 miles, how many minutes should it take him to walk 5 miles maintaining the same speed?

54. The perimeter of a 6-sided polygon is 56 cm. The length of three sides is 9 cm each. The length of two other sides is 8 cm each. What is the length of the missing side?

55. Convert $\frac{3}{25}$ to a decimal.

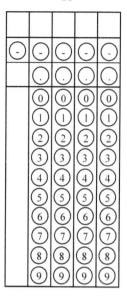

56. What is the value of $x^2 - 2xy + 2y^2$ when $x = 2, y = 3$?

57. If $4x - 3 = 5$, then $x =$

Answer Explanations #3

Editing/Revising

1. B: Choice *B* is the correct answer. Commas are only needed before the word *and* when it serves as a conjunction combining two complete sentences. In sentence 2, there are not two complete sentences being joined together, so no comma is needed after the word *ocean*.

2. A: Choice *A* is the correct answer. Book titles such as *The Outsiders* should be underlined or italicized. A comma is not needed before the prepositional phrase *in 1967*.

3. D: Choice *D* is the correct answer. Sentence 4 is a complex sentence, meaning it consists of one independent clause and one dependent clause. *While collecting such a large number of cans would be impossible for one person* is a dependent clause, so a comma is needed after the word *person*. The remainder of the sentence, *collecting them around your city might be an idea worth entertaining*, is an independent clause.

4. D: Choice *D* combines the two sentences correctly, including all of the information and clarifying the relationship between the idea that back-to-school shopping can be very costly and the idea that Texas hosts a tax-free weekend to help parents with the expense. Choices *A, B,* and *C* are incorrect because each sentence focuses on parents with multiple children, which is not the main idea of the first sentence.

5. B: Choice *B* is the correct answer. The verb in this sentence should be past tense because the events in the sentence happened previously. The word *comes* should be *came,* and the word *is* should be *was*.

6. C: Choice *C* is the correct answer. This sentence combines all of the information provided into one cohesive sentence.

7. D: Choice *D* is the correct answer. In this sentence, the word *some* indicates more than one, so the singular word *instance* needs to be *instances*. The plural form of *injury* is spelled correctly by replacing the *y* with *ies* (*injuries*).

8. D: Choice *D* is the correct answer. The word *because* is a subordinating conjunction that introduces the phrase, *Because giraffes eat over 100 pounds of twigs and leaves a day.* This dependent clause needs to have a comma after the word *day*.

9. A: Choice *A* is the correct answer. The transition phrase, *On this day,* helps to clarify that things are closed for a specific reason versus being closed in general.

10. A: Choice *A* is the correct answer. In sentence 1, the writer mentions Memorial Day and then Veterans Day in the statement of the main idea. Thus, the supporting sentences in the paragraph should also be in this order.

11. D: Choice *D* is the correct answer. This sentence rewords the writer's original sentence while still including all of the correct information. Choices *A, B,* and *C* either remain too wordy, provide inaccurate information, or fail to clarify the writer's idea.

Reading Comprehension

1. B: Mr. Button's wife is about to have a baby. The passage begins by giving the reader information about traditional birthing situations. Then, we are told that Mr. and Mrs. Button decide to go against tradition to have their baby in a hospital. The next few passages are dedicated to letting the reader know how Mr. Button dresses and goes to the hospital to welcome his new baby. There is a doctor in this excerpt, as Choice *C* indicates, and Mr. Button does put on clothes, as Choice *D* indicates. However, Mr. Button is not going to the doctor's office nor is he about to go shopping for new clothes.

2. A: The tone of the above passage is nervous and excited. We are told in the fourth paragraph that Mr. Button "arose nervously." We also see him running without caution to the doctor to find out about his wife and baby—this indicates his excitement. We also see him stuttering in a nervous yet excited fashion as he asks the doctor if it's a boy or girl. Though the doctor may seem a bit abrupt at the end, indicating a bit of anger or shame, neither of these choices is the overwhelming tone of the entire passage.

3. C: Dedicated. Mr. Button is dedicated to the task before him. Choice *A*, numbed, Choice *B*, chained, and Choice *D*, moved, all could grammatically fit in the sentence. However, they are not synonyms with *consecrated* like Choice *C* is.

4. D: Giving readers a visual picture of what the doctor is doing. The author describes a visual image—the doctor rubbing his hands together—first and foremost. The author may be trying to make a comment about the profession; however, the author does not "explain the detail of the doctor's profession" as Choice *B* suggests.

5. D: To introduce the setting of the story and its characters. We know we are being introduced to the setting because we are given the year in the very first paragraph along with the season: "one day in the summer of 1860." This is a classic structure of an introduction of the setting. We are also getting a long explanation of Mr. Button, what his work is, who is related to him, and what his life is like in the third paragraph.

6. B: "Talk sense!" is an example of an imperative sentence. An imperative sentence gives a command. The doctor is commanding Mr. Button to talk sense. Choice *A* is an example of an exclamatory sentence, which expresses excitement. Choice *C* is an example of an interrogative sentence—these types of sentences ask questions. Choice *D* is an example of a declarative sentence. This means that the character is simply making a statement.

7. C: The point of view is told in third person omniscient. We know this because the story starts out with us knowing something that the character does not know: that her husband has died. Mrs. Mallard eventually comes to know this, but we as readers know this information before it is broken to her. In third person limited, Choice *D*, we would only see and know what Mrs. Mallard herself knew, and we would find out the news of her husband's death when she found out the news, not before.

8. A: The way Mrs. Mallard reacted to her husband's death. The irony in this story is called situational irony, which means the situation that takes place is different than what the audience anticipated. At the beginning of the story, we see Mrs. Mallard react with a burst of grief to her husband's death. However, once she's alone, she begins to contemplate her future and says the word "free" over and over. This is quite a different reaction from Mrs. Mallard than what readers expected from the first of the story.

9. B: The word "elusive" most closely means "indefinable." Horrible, Choice *A*, doesn't quite fit with the tone of the word "subtle" that comes before it. Choice *C*, "quiet," is more closely related to the word "subtle." Choice *D*, "joyful," also doesn't quite fit the context here. "Indefinable" is the best option.

10. D: Mrs. Mallard, a newly widowed woman, finds unexpected relief in her husband's death. A summary is a brief explanation of the main point of a story. The story mostly focuses on Mrs. Mallard and her reaction to her husband's death, especially in the room when she's alone and contemplating the present and future. All of the other answer choices except Choice *C* are briefly mentioned in the story; however, they are not the main focus of the story.

11. D: The interesting thing about this story is that feelings that are confused, joyful, and depressive all play a unique and almost equal part of this story. There is no one right answer here, because the author seems to display all of these emotions through the character of Mrs. Mallard. She displays feelings of depressiveness by her grief at the beginning; then, when she receives feelings of joy, she feels moments of confusion. We as readers cannot help but go through these feelings with the character. Thus, the author creates a tone of depression, joy, and confusion, all in one story.

12. C: The word "tumultuously" most nearly means "violently." Even if you don't know the word "tumultuously," look at the surrounding context to figure it out. The next few sentences we see Mrs. Mallard striving to "beat back" the "thing that was approaching to possess her." We see a fearful and almost violent reaction to the emotion that she's having. Thus, her chest would rise and fall turbulently, or violently.

13. D: Extended metaphor. Metaphor is a direct comparison between two things, and extended metaphor is a lengthy, well-developed metaphor that usually extends over the length of the poem. In this poem, Keats forms an extended metaphor by drawing a comparison between the four seasons of nature and the "seasons" that humans experience from youth to old age.

14. A: Ponder. This question can be answered using context clues from the sentence: "Spring's honied cud of youthful thought he loves / To ruminate, and by such dreaming high / Is nearest unto heaven." Following the word "ruminate," it's restated as "such dreaming"; also, immediately before is the expression "youthful thought." Together, this sentence describes a young man pleasantly daydreaming. The only word related to thinking and daydreaming is "ponder," Choice *A*.

15. C: He spends less time thinking about beautiful things. This is a general comprehension question. The narrator describes a man in Autumn "contented so . . . to let fair things / Pass by unheeded." In this case, "fair" is another word for "beautiful," and letting things "pass by unheeded" means "he doesn't pay attention to them." In contrast, a man in the Spring and Summer of life spends time appreciating and daydreaming about beautiful things.

16. A: Winter represents the end of man's life. This is a purpose question, but it also requires readers to understand that this poem is an extended metaphor. Since the narrator is developing an extended comparison between seasons and life, it's natural that winter should come last because it's the season of death, dormancy, and "pale" nature (unlike, say, Spring, which is a season of life and rebirth in nature).

17. C: People change as they move through different stages of life. This is an inference question asking readers to understand the narrator's perspective. Choices *B* and *D* both include an opinion or advice to the reader, while the tone of the poem is more neutral or purely descriptive (the narrator is simply describing the stages of life, rather than advising readers on how to behave). Choice *C* more closely agrees with the comparison that the narrator sets up in the poem; just as seasons change in nature, people also change throughout their lives.

18. B: He would postpone or avoid death. This is both a vocabulary and a comprehension question. Based on the poem's extended metaphor, readers can gather that Winter is a metaphor for the end of life; all people must pass through Winter or else they would never die. Looking at the poem's vocabulary, "mortal" refers to human's limited life span (the opposite of "immortal"), and "forego" means to turn something down.

19. A: "in the mind of man" (2). This is a fairly straightforward question about literary devices. Alliteration refers to repetition of a word's beginning sound, and Choice *A* is the only example of that ("mind" and "man" both start with the letter M).

20. D: To define and describe examples of spinoff technology. This is a purpose question—*why* did the author write this? The article contains facts, definitions, and other objective information without telling a story or arguing an opinion. In this case, the purpose of the article is to inform the reader. The only answer choice related to giving information is Choice *D*: to define and describe.

21. A: A general definition followed by more specific examples. This organization question asks readers to analyze the structure of the essay. The topic of the essay is spinoff technology; the first paragraph gives a general definition of the concept, while the following two paragraphs offer more detailed examples to help illustrate this idea.

22. C: They were looking for ways to add health benefits to food. This reading comprehension question can be answered based on the second paragraph—scientists were concerned about astronauts' nutrition and began researching nutritional supplements. Choice *A* isn't true because it reverses the order of discovery (first NASA identified algae for astronaut use, and then it was further developed for use in baby food).

23. B: Related to the brain. This vocabulary question could be answered based on the reader's prior knowledge, but the passage provides context clues for readers who've never encountered the word "neurological." The next sentence talks about "this algae's potential to boost brain health," which is a paraphrase of "neurological benefits." From this context, readers should be able to infer that "neurological" relates to the brain.

24. D: To give an example of valuable space equipment. This purpose question requires readers to understand the relevance of the given detail. In this case, the author mentions "costly and crucial equipment" before space suit visors, which are given as an example of something valuable. Choice *A* isn't correct because fashion is only related to sunglasses, not to NASA equipment. Choice *B* can be eliminated because it's simply not mentioned. While Choice *C* seems like it could be true, it's not relevant.

25. C: It's difficult to make money from scientific research. The article gives several examples of how businesses have capitalized on NASA research, so it's unlikely that the author would agree with this statement. Evidence for the other answer choices can be found in the article: In Choice *A*, the author mentions that "many consumers are unaware that products they are buying are based on NASA research"; Choice *B* is a general definition of spinoff technology; and Choice *D* is mentioned in the final paragraph.

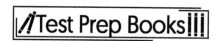

26. B: Narrative, Choice *A*, means a written account of connected events. Think of narrative writing as a story. Choice *C*, expository writing, generally seeks to explain or describe some phenomena, whereas Choice *D*, technical writing, includes directions, instructions, and/or explanations. This passage is definitely persuasive writing, which hopes to change someone's beliefs based on an appeal to reason or emotion. The author is aiming to convince the reader that smoking is terrible. They use health, price, and beauty in their argument against smoking, so Choice *B*, persuasive, is the correct answer.

27. B: The author is clearly opposed to tobacco. He cites disease and deaths associated with smoking. He points to the monetary expense and aesthetic costs. Choice *A* is wrong because alternatives to smoking are not even addressed in the passage. Choice *C* is wrong because it does not summarize the passage; rather, it is just a premise. Choice *D* is wrong because, while these statistics do support the argument, they do not represent a summary of the piece. Choice *B* is the correct answer because it states the three critiques offered against tobacco and expresses the author's conclusion.

28. C: We are looking for something the author would agree with, so it will almost certainly be anti-smoking or an argument in favor of quitting smoking. Choice *A* is wrong because the author does not speak against means of cessation. Choice *B* is wrong because the author does not reference other substances but does speak of how addictive nicotine—a drug in tobacco—is. Choice *D* is wrong because the author certainly would not encourage reducing taxes to encourage a reduction of smoking costs, thereby helping smokers to continue the habit. Choice *C* is correct because the author is definitely attempting to persuade smokers to quit smoking.

29. D: Here, we are looking for an opinion of the author's rather than a fact or statistic. Choice *A* is wrong because quoting statistics from the Centers of Disease Control and Prevention is stating facts, not opinions. Choice *B* is wrong because it expresses the fact that cigarettes sometimes cost more than a few gallons of gas. It would be an opinion if the author said that cigarettes were not affordable. Choice *C* is incorrect because yellow stains are a known possible adverse effect of smoking. Choice *D* is correct as an opinion because smell is subjective. Some people might like the smell of smoke, they might not have working olfactory senses, and/or some people might not find the smell of smoke akin to "pervasive nastiness," so this is the expression of an opinion. Thus, Choice *D* is the correct answer.

30. D: Criticize a theory by presenting counterevidence. The author mentions anti-Stratfordian arguments in the first paragraph, but then goes on to debunk these theories with facts about Shakespeare's life in the second and third paragraphs. Choice *A* is incorrect because the author is far from unbiased; in fact, the author clearly disagrees with anti-Stratfordians. Choice *B* is also incorrect because it's more closely aligned with the beliefs of anti-Stratfordians. Choice *C* can be eliminated because, while it's true that the author gives historical background, the purpose is using that information to disprove a theory.

31. B: "But in fact, there isn't much substance to such speculation, and most anti-Stratfordian arguments can be refuted with a little background about Shakespeare's time and upbringing." The thesis is a statement that contains the author's topic and main idea. As seen in the previous question, the purpose of this article is to use historical evidence to provide counterarguments to anti-Stratfordians. Choice *A* is simply a definition; Choice *C* is a supporting detail, not a main idea; and Choice *D* represents an idea of anti-Stratfordians, not the author's opinion.

32. B: By explaining grade school curriculum in Shakespeare's time. This question asks readers to refer to the organizational structure of the article and demonstrate understanding of how the author provides details to support the argument. This particular detail can be found in the second paragraph: "even though he did not attend university, grade school education in Shakespeare's time was actually quite rigorous."

33. D: In Shakespeare's time, glove-makers weren't part of the upper class. Anti-Stratfordians doubt Shakespeare's ability because he wasn't from the upper class; his father was a glove-maker; therefore, in at least this example, glove-makers weren't included in the upper class. This is an example of inductive reasoning, using two specific pieces of information to draw a more general conclusion.

34. C: It's an example of a play set outside of England. This detail comes from the third paragraph, where the author responds to skeptics who claim that Shakespeare wrote too much about places he never visited, so *Romeo and Juliet* is mentioned as a famous example of a play with a foreign setting. In order to answer this question, readers need to understand the author's purpose in the third paragraph and how the author uses details to support this purpose. Choices *A* and *D* aren't mentioned, and Choice *B* is clearly false because the passage mentions more than once that Shakespeare never left England.

35. A: It's possible to learn things from reading rather than firsthand experience. This inference can be made from the final paragraph, where the author refutes anti-Stratfordian skepticism by noting that books about life in Europe could circulate throughout London. From this statement, readers can conclude the author believes it's possible that Shakespeare learned about European culture from books. Choice *B* isn't true because the author believes that Shakespeare contributed to English literature without traveling extensively. Similarly, Choice *C* isn't a good answer because the author explains how Shakespeare got his education without attending a university. Choice *D* can also be eliminated because the author describes Shakespeare's genius, and Shakespeare clearly isn't from Italy.

36. A: The purpose is to inform the reader about what assault is and how it is committed. Choice *B* is incorrect because the passage does not state that assault is a lesser form of lethal force, only that an assault can use lethal force, or alternatively, lethal force can be utilized to counter a dangerous assault. Choice *C* is incorrect because the passage is informative and does not have a set agenda. Finally, Choice *D* is incorrect because although the author uses an example in order to explain assault, it is not indicated that this is the author's personal account.

37. C: If the man being attacked in an alley by another man with a knife used self-defense by lethal force, it would not be considered illegal. The presence of a deadly weapon indicates mal-intent and because the individual is isolated in an alley, lethal force in self-defense may be the only way to preserve his life. Choices *A* and *B* can be ruled out because in these situations, no one is in danger of immediate death or bodily harm by someone else. Choice *D* is an assault and does exhibit intent to harm, but this situation isn't severe enough to merit lethal force; there is no intent to kill.

38. B: As discussed in the second passage, there are several forms of assault, like assault with a deadly weapon, verbal assault, or threatening posture or language. Choice *A* is incorrect because the author does mention what the charges are on assaults; therefore, we cannot assume that they are more or less than unnecessary use of force charges. Choice *C* is incorrect because anyone is capable of assault; the author does not state that one group of people cannot commit assault. Choice *D* is incorrect because assault is never justified. Self-defense resulting in lethal force can be justified.

39. D: The use of lethal force is not evaluated on the intent of the user, but rather on the severity of the primary attack that warranted self-defense. This statement most undermines the last part of the passage because it directly contradicts how the law evaluates the use of lethal force. Choices *A* and *B* are stated in the paragraph, so they do not undermine the explanation from the author. Choice *C* does not necessarily undermine the passage, but it does not support the passage either. It is more of an opinion that does not offer strength or weakness to the explanation.

40. C: An assault with deadly intent can lead to an individual using lethal force to preserve their well-being. Choice *C* is correct because it clearly establishes what both assault and lethal force are and gives the specific way in which the two concepts meet. Choice *A* is incorrect because lethal force doesn't necessarily result in assault. This is also why Choice *B* is incorrect. Not all assaults would necessarily be life-threatening to the point where lethal force is needed for self-defense. Choice *D* is compelling but ultimately too vague; the statement touches on aspects of the two ideas but fails to present the concrete way in which the two are connected to each other.

41. A: Both passages open by defining a legal concept and then continue to describe situations in order to further explain the concept. Choice *D* is incorrect because while the passages utilize examples to help explain the concepts discussed, the author doesn't indicate that they are specific court cases. It's also clear that the passages don't open with examples, but instead, they begin by defining the terms addressed in each passage. This eliminates Choice *B*, and ultimately reveals Choice *A* to be the correct answer. Choice *A* accurately outlines the way both passages are structured. Because the passages follow a nearly identical structure, the Choice *C* can easily be ruled out.

42. C: *Extraneous* most nearly means *superfluous*, or *trivial*. Choice *A*, *indispensable*, is incorrect because it means the opposite of *extraneous*. Choice *B*, *bewildering*, means *confusing* and is not relevant to the context of the sentence. Finally, Choice *D* is wrong because although the prefix of the word is the same, *ex-*, the word *exuberant* means *elated* or *enthusiastic*, and is irrelevant to the context of the sentence.

43. A: The author's purpose is to bring to light an alternative view on human perception by examining the role of technology in human understanding. This is a challenging question because the author's purpose is somewhat open-ended. The author concludes by stating that the questions regarding human perception and observation can be approached from many angles. Thus, the author does not seem to be attempting to prove one thing or another. Choice *B* is incorrect because we cannot know for certain whether the electron experiment is the latest discovery in astroparticle physics because no date is given. Choice *C* is a broad generalization that does not reflect accurately on the writer's views. While the author does appear to reflect on opposing views of human understanding (Choice *D*), the best answer is Choice *A*.

44. C: It presents a problem, explains the details of that problem, and then ends with more inquiry. The beginning of this paragraph literally "presents a conundrum," explains the problem of partial understanding, and then ends with more questions, or inquiry. There is no solution offered in this paragraph, making Choices *A* and *B* incorrect. Choice *D* is incorrect because the paragraph does not begin with a definition.

45. B: The electrons passed through both holes and then onto the plate. Choices *A* and *C* are wrong because such movement is not mentioned at all in the text. In the passage the author says that electrons that were physically observed appeared to pass through one hole or another. Remember, the electrons that were observed doing this were described as acting like particles. Therefore, Choice *D* is wrong. Recall that the plate actually recorded electrons passing through both holes simultaneously and hitting the plate. This behavior, the electron activity that wasn't seen by humans, was characteristic of waves. Thus, Choice *B* is the right answer.

46. C: The author uses "gravity" to demonstrate an example of natural phenomena humans discovered and understand without the use of tools or machines. Choice *A* mirrors the language in the beginning of the paragraph but is incorrect in its intent. Choice *B* is incorrect; the paragraph mentions nothing of "not knowing the true nature of gravity." Choice *D* is incorrect as well. There is no mention of an "alternative solution" to new technology in this paragraph.

Math

1. B: 300 miles in 4 hours is $300 \div 4 = 75$ miles per hour. In 1.5 hours, the car will go 1.5×75 miles, or 112.5 miles.

2. C: One apple/orange pair costs $3 total. Therefore, Jan bought $90 \div 3 = 30$ total pairs, and hence, she bought 30 oranges.

3. C: The formula for the volume of a box with rectangular sides is the length times width times height, so:

$$5 \times 6 \times 3 - 90 \text{ cubic feet}$$

4. D: First, the train's journey in the real word is $3 \times 50 = 150$ miles. On the map, 1 inch corresponds to 10 miles, so there is $150 \div 10 = 15$ inches on the map.

5. B: The total trip time is:

$$1 + 3.5 + 0.5 = 5 \text{ hours}$$

The total time driving is:

$$1 + 0.5 = 1.5 \text{ hours}$$

So, the fraction of time spent driving is:

$$\frac{1.5}{5}$$

or

$$\frac{3}{10}$$

To get the percentage, convert this to a fraction out of 100. The numerator and denominator are multiplied by 10, with a result of $\frac{30}{100}$. The percentage is the numerator in a fraction out of 100, so 30%.

6. B: The formula for the volume of a cylinder is $\pi r^2 h$, where r is the radius and h is the height. The diameter is twice the radius, so these barrels have a radius of 1 foot. That means each barrel has a volume of:

$$\pi \times 1^2 \times 3 = 3\pi \text{ cubic feet}$$

Since there are three of them, the total is $3 \times 3\pi = 9\pi$ cubic feet.

7. A: The tip is not taxed, so he pays 5% tax only on the $10. 5% of $10 is:

$$0.05 \times 10 = \$0.50$$

Add up $10 + $2 + $0.50 to get $12.50.

8. B: To simplify the given equation, the first step is to make all exponents positive by moving them to the opposite place in the fraction. This expression becomes:

$$\frac{4b^3 b^2}{a^1 a^4} \times \frac{3a}{b}$$

Then the rules for exponents can be used to simplify. Multiplying the same bases means the exponents can be added. Dividing the same bases means the exponents are subtracted.

9. A: To find 20% of 40, simply multiply 40 by .20. This will give you the answer of 8.

10. A: The value went up by:

$$\$165,000 - \$150,000 = \$15,000$$

Out of $150,000, this is:

$$\frac{15,000}{150,000} = \frac{1}{10}$$

Convert this to having a denominator of 100, the result is $\frac{10}{100}$ or 10%.

11. D: The total faculty is $15 + 20 = 35$. Therefore, the faculty to student ratio is 35:200. Then, to simplify this ratio, both the numerator and the denominator are divided by 5, since 5 is a common factor of both, which yields 7:40.

12. B: The journey will be $5 \times 3 = 15$ miles. A car traveling at 60 miles per hour is traveling at 1 mile per minute. So, it will take $15 \div 1 = 15$ minutes to take the journey.

13. A: Taylor's total income is:

$$\$20,000 + \$10,000 = \$30,000$$

15% of this is:

$$\frac{15}{100} = \frac{3}{20}$$

$$\frac{3}{20} \times \$30,000 = \frac{90,000}{20} = \frac{9000}{2} = \$4500$$

14. C: The surface area formula for a rectangular prism is $2(lw + lh + wh)$. If the dimensions are converted to decimals, 3.5 ft can be used for length, 2.5 for width, and 1.5 for height. Substituting these values into the formula yields:

$$2(3.5 \times 2.5 + 3.5 \times 1.5 + 2.5 \times 1.5)$$

The surface area of the box is 35.5 ft^2.

15. D: Kristen bought four DVDs, which would cost a total of:

$$4 \times 15 = \$60$$

She spent a total of $100, so she spent:

$$\$100 - \$60 = \$40 \text{ on CDs}$$

Since they cost $10 each, she must have purchased:

$$40 \div 10 = 4 \text{ CDs}$$

16. D: This problem can be solved by setting up a proportion involving the given information and the unknown value. The proportion is:

$$\frac{21 \; pages}{4 \; nights} = \frac{140 \; pages}{x \; nights}$$

Solving the proportion by cross-multiplying, the equation becomes $21x = 4 \times 140$, where $x = 26.67$. Since it is not an exact number of nights, the answer is rounded up to 27 nights. Twenty-six nights would not give Sarah enough time.

17. D: This problem can be solved by using unit conversion. The initial units are miles per minute. The final units need to be feet per second. Converting miles to feet uses the equivalence statement 1 mile = 5,280 feet. Converting minutes to seconds uses the equivalence statement 1 minute = 60 seconds. Setting up the ratios to convert the units is shown in the following equation:

$$\frac{72 \; miles}{90 \; minutes} \times \frac{1 \; minute}{60 \; seconds} \times \frac{5280 \; feet}{1 \; mile} = 70.4 \; ft. per \; second$$

The initial units cancel out, and the new units are left.

18. C: The sum total percentage of a pie chart must equal 100%. Since the CD sales take up less than half of the chart and more than a quarter (25%), it can be determined to be 40% overall. This can also be measured with a protractor. The angle of a circle is 360°. Since 25% of 360 would be 90° and 50% would be 180°, the angle percentage of CD sales falls in between; therefore, it would be Choice *C*.

19. B: Since $850 is the price *after* a 20% discount, $850 represents 80% of the original price. To determine the original price, set up a proportion with the ratio of the sale price (850) to original price (unknown) equal to the ratio of sale percentage:

$$\frac{850}{x} = \frac{80}{100}$$

(where *x* represents the unknown original price)

To solve a proportion, cross multiply the numerators and denominators and set the products equal to each other:

$$850 \times 100 = 80x$$

Multiplying each side results in the equation $85,000 = 80x$.

To solve for *x*, both sides get divided by 80: $\frac{85,000}{80} = \frac{80x}{80}$, resulting in $x = 1062.5$. Remember that *x* represents the original price. Subtracting the sale price from the original price $1062.50 - \$850$ indicates that Frank saved $212.50.

20. C: Scientific notation division can be solved by grouping the first terms together and grouping the tens together. The first terms can be divided, and the tens terms can be simplified using the rules for exponents. The initial expression becomes 0.4×10^4. This is not in scientific notation because the first number is not between 1 and 10. Shifting the decimal and subtracting one from the exponent, the answer becomes 4.0×10^3.

21. A: The relative error can be found by finding the absolute error and making it a percent of the true value. The absolute value is:

$$36 - 35.75 = 0.25$$

This error is then divided by 35.75—the true value—to find 0.7%.

22. C: The sample space is made up of:

$$8 + 7 + 6 + 5 = 26 \text{ balls}$$

The probability of pulling each individual ball is $\frac{1}{26}$. Since there are 7 yellow balls, the probability of pulling a yellow ball is $\frac{7}{26}$.

23. B: For the first card drawn, the probability of a King being pulled is $\frac{4}{52}$. Since this card isn't replaced, if a King is drawn first the probability of a King being drawn second is $\frac{3}{51}$. The probability of a King being drawn in both the first and second draw is the product of the two probabilities: $\frac{4}{52} \times \frac{3}{51} = \frac{12}{2652}$ which, divided by 12, equals $\frac{1}{221}$.

24. C: $\angle A$ and $\angle B$ are supplementary angles, so together they equal $180°$. The value of x must be calculated first. The equation becomes:

$$2x + (4x - 6) = 180$$

Combining like terms yields $6x = 186$ so $x = 31$. Plugging the value of x into the equation for $\angle B$ results in:

$$4(31) - 6 = 118°$$

25. D: The addition rule is necessary to determine the probability because a 6 can be rolled on either roll of the die. The rule used is:

$$P(A \text{ or } B) = P(A) + P(B) - P(A \text{ and } B)$$

The probability of a 6 being individually rolled is $\frac{1}{6}$ and the probability of a 6 being rolled twice is:

$$\frac{1}{6} \times \frac{1}{6} = \frac{1}{36}$$

Therefore, the probability that a 6 is rolled at least once is:

$$\frac{1}{6} + \frac{1}{6} - \frac{1}{36} = \frac{11}{36}$$

26. C: 85% of a number means multiplying that number by 0.85. So:

$$0.85 \times 20 = \frac{85}{100} \times \frac{20}{1}$$

which can be simplified to:

$$\frac{17}{20} \times \frac{20}{1} = 17$$

27. A: To find the fraction of the bill that the first three people pay, the fractions need to be added, which means finding common denominator. The common denominator will be 60.

$$\frac{1}{5} + \frac{1}{4} + \frac{1}{3} = \frac{12}{60} + \frac{15}{60} + \frac{20}{60} = \frac{47}{60}$$

The remainder of the bill is:

$$1 - \frac{47}{60} = \frac{60}{60} - \frac{47}{60} = \frac{13}{60}$$

28. B: 30% is $\frac{3}{10}$. The number itself must be $\frac{10}{3}$ of 6, or $\frac{10}{3} \times 6 = 10 \times 2 = 20$.

29. A: Changing these numbers to improper fractions yields: $\frac{11}{3} - \frac{9}{5}$. Take 15 as a common denominator:

$$\frac{11}{3} - \frac{9}{5}$$

$$\frac{55}{15} - \frac{27}{15} = \frac{28}{15} = 1\frac{13}{15}$$

(when rewritten to get rid of the partial fraction).

30. A: If each man gains 10 pounds, every original data point will increase by 10 pounds. Therefore, the man with the original median will still have the median value, but that value will increase by 10. The smallest value and largest value will also increase by 10 and, therefore, the difference between the two won't change. The range does not change in value and, thus, remains the same.

31. C: To find the average of a set of values, add the values together and then divide by the total number of values. In this case, include the unknown value of what Dwayne needs to score on his next test, in order to solve it.

$$\frac{78 + 92 + 83 + 97 + x}{5} = 90$$

Add the unknown value to the new average total, which is 5. Then multiply each side by 5 to simplify the equation, resulting in:

$$78 + 92 + 83 + 87 + x = 450$$

$$350 + x = 450$$

$$x = 100$$

Dwayne would need to get a perfect score of 100 in order to get an average of at least 90.

Test this answer by substituting back into the original formula.

$$\frac{78 + 92 + 83 + 97 + 100}{5} = 90$$

32. B: The first step is to calculate the difference between the larger value and the smaller value.

$$378 - 252 = 126$$

To calculate this difference as a percentage of the original value, and thus calculate the percentage *increase*, 126 is divided by 252, then this result is multiplied by 100 to find the percentage = 50%.

33. B: Multiplying by 10^{-3} means moving the decimal point three places to the left, putting in zeroes as necessary.

34. B: $\frac{5}{2} \div \frac{1}{3} = \frac{5}{2} \times \frac{3}{1} = \frac{15}{2} = 7.5$.

35. A: The solid dot is located between -2 and -3, and the open dot is located between 1 and 2. Therefore, x is between -2.5 and 1.5, which can be converted to $-\frac{5}{2}$ and $\frac{3}{2}$. The solid dot indicates greater than or equal to, and the open dot indicates less than so the inequality is:

$$-\frac{5}{2} \leq x < \frac{3}{2}$$

36. C: If she has used $\frac{1}{3}$ of the paint, she has $\frac{2}{3}$ remaining. $2\frac{1}{2}$ gallons are the same as $\frac{5}{2}$ gallons. The calculation is:

$$\frac{2}{3} \times \frac{5}{2} = \frac{5}{3} = 1\frac{2}{3} \text{ gallons}$$

37. B: To simplify this inequality, subtract 3 from both sides to get $-\frac{1}{2}x \geq -1$. Then, multiply both sides by -2 (remembering this flips the direction of the inequality) to get $x \leq 2$.

38. D: The slope is given by the change in y divided by the change in x. Specifically, it's:

$$slope = \frac{y_2 - y_1}{x_2 - x_1}$$

The first point is $(-5, -3)$ and the second point is $(0, -1)$. Work from left to right when identifying coordinates. Thus the point on the left is point 1 $(-5, -3)$, and the point on the right is point 2 $(0, -1)$.

Now we need to just plug those numbers into the equation:

$$slope = \frac{-1 - (-3)}{0 - (-5)}$$

It can be simplified to:

$$slope = \frac{-1+3}{0+5}$$

$$slope = \frac{2}{5}$$

39. B: The figure is composed of three sides of a square and a semicircle. The sides of the square are simply added: $8 + 8 + 8 = 24$ inches. The circumference of a circle is found by the equation $C = 2\pi r$. The radius is 4 in, so the circumference of the circle is 25.13 in. Only half of the circle makes up the outer border of the figure (part of the perimeter) so half of 25.13 in is 12.565 in. Therefore, the total perimeter is:

$$24\ in + 12.565\ in = 36.565\ in$$

The other answer choices use the incorrect formula or fail to include all of the necessary sides.

40. A: The first step is to determine the unknown, which is in terms of the length, l.

The second step is to translate the problem into the equation using the perimeter of a rectangle:

$$P = 2l + 2w$$

The width is the length minus 2 centimeters. The resulting equation is:

$$2l + 2(l - 2) = 44$$

The equation can be solved as follows:

$2l + 2l - 4 = 44$	Apply the distributive property on the left side of the equation
$4l - 4 = 44$	Combine like terms on the left side of the equation
$4l = 48$	Add 4 to both sides of the equation
$l = 12$	Divide both sides of the equation by 4

The length of the rectangle is 12 centimeters. The width is the length minus 2 centimeters, which is 10 centimeters. Checking the answers for length and width forms the following equation:

$$44 = 2(12) + 2(10)$$

The equation can be solved using the order of operations to form a true statement: $44 = 44$.

41. B: $3x^2 - 3x + 11$. By distributing the implied one in front of the first set of parentheses and the -1 in front of the second set of parentheses, the parenthesis can be eliminated:

$$1(5x^2 - 3x + 4) - 1(2x^2 - 7) = 5x^2 - 3x + 4 - 2x^2 + 7$$

Next, like terms (same variables with same exponents) are combined by adding the coefficients and keeping the variables and their powers the same:

$$5x^2 - 3x + 4 - 2x^2 + 7 = 3x^2 - 3x + 11$$

42. D: Three girls for every two boys can be expressed as a ratio: 3:2. This can be visualized as splitting the school into 5 groups: 3 girl groups and 2 boy groups. The number of students that are in each group can be found by dividing the total number of students by 5:

650 divided by 5 equals 1 part, or 130 students per group

To find the total number of girls, the number of students per group (130) is multiplied by how the number of girl groups in the school (3). This equals 390, Choice *D*.

43. C: Kimberley worked 4.5 hours at the rate of $10/h and 1 hour at the rate of $12/h. The problem states that her pay is rounded to the nearest hour, so the 4.5 hours would round up to 5 hours at the rate of $10/h.

$$5 \times \$10 + 1 \times \$12$$

$$\$50 + \$12 = \$62$$

44. C: The average is calculated by adding all six numbers, then dividing by 6. The first five numbers have a sum of 25. If the total divided by 6 is equal to 6, then the total itself must be 36. The sixth number must be $36 - 25 = 11$.

45. B: To find the percentage of cars that are not black or white, the total number of cars must be found first. The total number of cars is 191, and the total of all the cars that are not black or white is 87. The percentage can be found using the following calculation:

$$\frac{87}{191} = 0.455 = 46\%$$

46. A: This problem can be solved by simple multiplication and addition. Since the sale date is over six years apart, 6 can be multiplied by 12 for the number of months in a year, and then the remaining 4 months can be added.

$$(6 \times 12) + 4 = ?$$

$$72 + 4 = 76$$

47. D: This problem can be solved using basic arithmetic. Xavier starts with 20 apples, then gives his sister half, so 20 divided by 2.

$$\frac{20}{2} = 10$$

He then gives his neighbor 6, so 6 is subtracted from 10.

$$10 - 6 = 4$$

Lastly, he uses ¾ of his apples to make an apple pie, so to find remaining apples, the first step is to subtract ¾ from one and then multiply the difference by 4.

$$\left(1 - \frac{3}{4}\right) \times 4 = ?$$

$$\left(\frac{4}{4} - \frac{3}{4}\right) \times 4 = ?$$

$$\left(\frac{1}{4}\right) \times 4 = 1$$

48. C: Nothing is added to x and y since the center is 0 and 5^2 is 25. Choice A is not the correct answer because you do not subtract the radius from x and y. Choice B is not the correct answer because you must square the radius on the right side of the equation. Choice D is not the correct answer because you do not add the radius to x and y in the equation.

49. B: To solve this correctly, keep in mind the order of operations with the mnemonic PEMDAS (Please Excuse My Dear Aunt Sally). This stands for Parentheses, Exponents, Multiplication, Division, Addition, Subtraction. Taking it step by step, solve the parentheses first:

$$4 \times 7 + 4^2 \div 2$$

Then, apply the exponent:

$$4 \times 7 + 16 \div 2$$

Multiplication and division are both performed next:

$$28 + 8 = 36$$

50. A: To calculate the range in a set of data, subtract the lowest value from the highest value. In this graph, the range of Mr. Lennon's students is 5, which can be seen physically in the graph as having the smallest difference between the highest value and the lowest value compared with the other teachers.

51. C: The volume of a cylinder is $\pi r^2 h$, and $\pi \times 6^2 \times 2$ is $72\,\pi$ cm³. Choice A is not the correct answer because that is only $6^2 \times \pi$. Choice B is not the correct answer because that is $2^2 \times 6 \times \pi$. Choice D is not the correct answer because that is $2^3 \times 6 \times \pi$.

52. B: This answer is correct because $3^2 + 4^2$ is $9 + 16$, which is 25. Taking the square root of 25 is 5. Choice A is not the correct answer because that is $3 + 4$. Choice C is not the correct answer because that is stopping at $3^2 + 4^2$ is $9 + 16$, which is 25. Choice D is not the correct answer because that is 3×4.

53.

80. To solve the problem, a proportion is written consisting of ratios comparing distance and time. One way to set up the proportion is:

$$\frac{3}{48} = \frac{5}{x} \left(\frac{distance}{time} = \frac{distance}{time} \right)$$

where *x* represents the unknown value of time. To solve a proportion, the ratios are cross-multiplied:

$$(3)(x) = (5)(48) \rightarrow 3x = 240$$

The equation is solved by isolating the variable, or dividing by 3 on both sides, to produce *x* = 80.

54.

			1	3
⊖	⊖	⊖	⊖	⊖
	·	·	·	·
	0	0	0	0
	1	1	●	1
	2	2	2	2
	3	3	3	●
	4	4	4	4
	5	5	5	5
	6	6	6	6
	7	7	7	7
	8	8	8	8
	9	9	9	9

13. Perimeter is found by calculating the sum of all sides of the polygon. $9 + 9 + 9 + 8 + 8 + s = 56$, where s is the missing side length. Therefore, 43 plus the missing side length is equal to 56. The missing side length is 13 cm.

55.

0	.	1	2
⊖	⊖	⊖	⊖	⊖
	●	·	·
●	0	0	0
1	1	●	1
2	2	2	●
3	3	3	3
4	4	4	4
5	5	5	5
6	6	6	6
7	7	7	7
8	8	8	8
9	9	9	9

0.12. The fraction is converted so that the denominator is 100 by multiplying the numerator and denominator by 4, to get $\frac{3}{25} = \frac{12}{100}$. Dividing a number by 100 just moves the decimal point two places to the left, with a result of 0.12.

56.

			1	0

10. Each instance of x is replaced with a 2, and each instance of y is replaced with a 3 to get:

$$2^2 - 2 \times 2 \times 3 + 2 \times 3^2$$

$$4 - 12 + 18 = 10$$

57.

				2

2. Add 3 to both sides to get $4x = 8$. Then divide both sides by 4 to get $x = 2$.

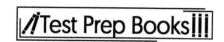

Dear SHSAT Test Taker,

We would like to start by thanking you for purchasing this practice test book for your SHSAT exam. We hope that we exceeded your expectations.

We strive to make our practice questions as similar as possible to what you will encounter on test day. With that being said, if you found something that you feel was not up to your standards, please send us an email and let us know.

We would also like to let you know about other books in our catalog that may interest you.

SAT

This can be found on Amazon: amazon.com/dp/1628456396

ACCUPLACER

amazon.com/dp/1628456515

SAT Math 1

amazon.com/dp/1628454717

We have study guides in a wide variety of fields. If the one you are looking for isn't listed above, then try searching for it on Amazon or send us an email.

Thanks Again and Happy Testing!
Product Development Team
info@studyguideteam.com

Interested in buying more than 10 copies of our product? Contact us about bulk discounts:

bulkorders@studyguideteam.com

FREE Test Taking Tips DVD Offer

To help us better serve you, we have developed a Test Taking Tips DVD that we would like to give you for FREE. **This DVD covers world-class test taking tips that you can use to be even more successful when you are taking your test.**

All that we ask is that you email us your feedback about your study guide. Please let us know what you thought about it – whether that is good, bad or indifferent.

To get your **FREE Test Taking Tips DVD**, email freedvd@studyguideteam.com with "FREE DVD" in the subject line and the following information in the body of the email:

    a. The title of your study guide.

    b. Your product rating on a scale of 1-5, with 5 being the highest rating.

    c. Your feedback about the study guide. What did you think of it?

    d. Your full name and shipping address to send your free DVD.

If you have any questions or concerns, please don't hesitate to contact us at freedvd@studyguideteam.com.

Thanks again!

CPSIA information can be obtained
at www.ICGtesting.com
Printed in the USA
BVHW010304070421
604344BV00008B/383

9 781628 457063